METHODS IN MOLECULAR BIOLOGY

D1807154

Series Editor
John M. Walker
School of Life and Medical Sciences
University of Hertfordshire
Hatfield, Hertfordshire, UK

For further volumes:
http://www.springer.com/series/7651

For over 35 years, biological scientists have come to rely on the research protocols and methodologies in the critically acclaimed *Methods in Molecular Biology* series. The series was the first to introduce the step-by-step protocols approach that has become the standard in all biomedical protocol publishing. Each protocol is provided in readily-reproducible step-by-step fashion, opening with an introductory overview, a list of the materials and reagents needed to complete the experiment, and followed by a detailed procedure that is supported with a helpful notes section offering tips and tricks of the trade as well as troubleshooting advice. These hallmark features were introduced by series editor Dr. John Walker and constitute the key ingredient in each and every volume of the *Methods in Molecular Biology* series. Tested and trusted, comprehensive and reliable, all protocols from the series are indexed in PubMed.

Macrophage Migration Inhibitory Factor

Methods and Protocols

Edited by

James Harris

Rheumatology Group, Centre for Inflammatory Diseases, Department of Medicine, School of Clinical Sciences at Monash Health, Faculty of Medicine, Nursing and Health Sciences, Monash Medical Centre, Monash University, Clayton, VIC, Australia

Eric F. Morand

Rheumatology Group, Centre for Inflammatory Diseases, Department of Medicine, School of Clinical Sciences at Monash Health, Faculty of Medicine, Nursing and Health Sciences, Monash Medical Centre, Monash University, Clayton, VIC, Australia

 Humana Press

Editors
James Harris
Rheumatology Group
Centre for Inflammatory Diseases
Department of Medicine
School of Clinical Sciences at Monash Health
Faculty of Medicine, Nursing and Health Sciences
Monash Medical Centre
Monash University
Clayton, VIC, Australia

Eric F. Morand
Rheumatology Group
Centre for Inflammatory Diseases
Department of Medicine
School of Clinical Sciences at Monash Health
Faculty of Medicine, Nursing and Health Sciences
Monash Medical Centre
Monash University
Clayton, VIC, Australia

ISSN 1064-3745   ISSN 1940-6029   (electronic)
Methods in Molecular Biology
ISBN 978-1-4939-9938-5   ISBN 978-1-4939-9936-1   (eBook)
https://doi.org/10.1007/978-1-4939-9936-1

This Humana imprint is published by the registered company Springer Science+Business Media, LLC, part of Springer Nature.
The registered company address is: 233 Spring Street, New York, NY 10013, U.S.A.

Preface

Macrophage migration inhibitor factor (MIF) is one of the most enigmatic molecules in biology. Highly conserved across species, including homologs in plants, its activity was first identified in the 1960s, in studies of cellular migration during delayed hypersensitivity responses. In two independent studies, a soluble factor inhibiting leukocyte migration in vitro was identified [1, 2]. Akin to many other proteins named in the early days of immunology, such as tumor necrosis factor (TNF), its range of functions vastly exceeds those implied by its historical name. Human and mouse MIF were not definitively cloned until the early 1990s [3, 4], after which the roles attributed to this intriguing molecule have grown and continue to do so. Indeed, studies now suggest that its name may be misleading, as it actually has chemotactic properties for a number of cell types, including monocytes, neutrophils, and lymphocytes [5, 6], recruiting and retaining them at sites of inflammation. Significantly, studies now highlight a wide repertoire of roles for MIF, both within and outside the immune system, that encompass both cytokine-like effects and discreet intracellular functions [7].

Indeed, the biology of MIF is an exemplar of the famous adage, attributed to both Aristotle and Albert Einstein, "the more we learn, the more we realise how little we know." The fact that the deletion of MIF results in large effects on cell biology and models of disease informs regarding its importance in pathology, and potential as a therapeutic target across a host of inflammatory and autoimmune diseases, including sepsis, arthritis, multiple sclerosis, systemic lupus erythematosus, and inflammatory bowel disease [8–12]. However, MIF-deficient mice are surprisingly healthy, consistent with substantial redundancy of MIF's effects with as yet unidentified other proteins. The fact that MIF is detectable in the extracellular space and that antibody neutralization has powerful inhibitory effects on immunity in vitro and in vivo imply cytokine-like function. However, MIF detectable in serum at high concentrations in health arguably indicates a nontraditional cytokine-like role. Moreover, its enzymatic actions, abundant intracellular reservoir, and multiple well-documented interactions with cytoplasmic proteins, point to important intracellular functions. Thus, there is still much to be discovered (and resolved) regarding the intracellular and extracellular actions of MIF and further studies will continue to take us closer to full realization of its therapeutic potential.

In addition to scientific curiosity, this therapeutic promise continues to drive research programs in MIF biology today. Repeated independent demonstrations of the potential benefit of MIF inhibition in inflammatory diseases is matched by strong evidence in multiple cancers. And yet, despite this evidence base and multiple attempts by both small biotech companies and major pharmaceutical companies to develop MIF inhibitors, no such treatment has reached the market. A lack of complete resolution of the biology of MIF is, we suggest, a major limiting factor on advancing this cause, as without knowledge of the biology, therapeutic inhibition of MIF carries a risk profile that is unknowable.

In this volume of *Methods in Molecular Biology*, we have sought to bring together expertise across a range of areas of biology in which MIF has been studied, to provide a methodological foundation for further MIF research. In each chapter, leading experts have outlined experimental approaches that can be replicated in new labs as interest in MIF research continues to grow. Methodological approaches to quantification of MIF in cell supernatants and human samples, using methods such as ELISA and Western blot analysis are described, along with detection of intracellular MIF using flow cytometry, and detection of MIF in the protozoan parasite *Entamoeba histolytica* using novel antibodies. We outline novel approaches to detect *Mif* expression in a marine invertebrate using rapid amplification of complementary DNA (cDNA) ends (RACE), a technique potentially applicable to a host of settings. We describe approaches to evaluating intracellular MIF expression, location, and interactions using co-IP and advanced imaging techniques, and the presence and location of the MIF receptor CD74 in tissues using immunohistochemistry. Assays to determine the role of MIF in cell migration and chemotaxis in vitro in vivo are described, as are assays to study the role of MIF in a wide range of physiological processes, including autophagy, inflammasome activation, cell death, osteoclastogenesis, and enteric permeability. Finally, with a view to translation of MIF biology to medicine, we describe approaches to the study of MIF in models of disease such as severe asthma, *Plasmodium* infection, and gout, along with methods for genotyping *MIF* polymorphisms in studies of human predisposition to disease.

We hope these works will facilitate future studies in MIF research, revealing new understanding of MIF biology in health and disease and leading to the realization of the significant translational promise of therapeutic MIF interventions.

Clayton, VIC, Australia   *James Harris*
                                                             Eric F. Morand

References

1. Bloom BR, Bennett B (1966) Mechanism of a reaction in vitro associated with delayed-type hypersensitivity. Science 153(3731):80–82
2. David JR (1966) Delayed hypersensitivity in vitro: its mediation by cell-free substances formed by lymphoid cell-antigen interaction. Proc Natl Acad Sci U S A 56(1):72–77
3. Bernhagen J et al (1993) MIF is a pituitary-derived cytokine that potentiates lethal endotoxaemia. Nature 365(6448):756–759
4. Bernhagen J et al (1994) Purification, bioactivity, and secondary structure analysis of mouse and human macrophage migration inhibitory factor (MIF). Biochemistry 33(47):14144–14155
5. Alampour-Rajabi S et al (2015) MIF interacts with CXCR7 to promote receptor internalization, ERK1/2 and ZAP-70 signaling, and lymphocyte chemotaxis. FASEB J 29(11):4497–4511
6. Bernhagen J et al (2007) MIF is a noncognate ligand of CXC chemokine receptors in inflammatory and atherogenic cell recruitment. Nat Med 13(5):587–596
7. Harris J et al (2019) Rediscovering MIF: new tricks for an old cytokine. Trends Immunol 40 (5):447–462
8. Benedek G et al (2017) MIF and D-DT are potential disease severity modifiers in male MS subjects. Proc Natl Acad Sci U S A 114(40):E8421–E8429
9. Calandra T et al (2000) Protection from septic shock by neutralization of macrophage migration inhibitory factor. Nat Med 6(2):164–170
10. de Jong YP et al (2001) Development of chronic colitis is dependent on the cytokine MIF. Nat Immunol 2(11):1061–1066
11. Lang T et al (2015) MIF: implications in the pathoetiology of systemic lupus erythematosus. Front Immunol 6:577
12. Morand EF, Leech M, Bernhagen J (2006) MIF: a new cytokine link between rheumatoid arthritis and atherosclerosis. Nat Rev Drug Discov 5(5):399–410

Contents

Contributors

JAWAD H. ABIDI • *Centre for Inflammatory Diseases, School of Clinical Sciences at Monash Health, Faculty of Medicine, Nursing and Health Sciences, Monash University, Clayton, VIC, Australia*

ELIZABETH H. AITKEN • *Department of Medicine (RMH), Peter Doherty Institute, University of Melbourne, Melbourne, VIC, Australia*

VENKATA SITA RAMA RAJU ALLAM • *Graduate School of Health, Faculty of Health, University of Technology Sydney, Ultimo, NSW, Australia*

FLAVIO ALMEIDA AMARAL • *Departamento de Bioquímica e Imunologia, Instituto de Ciências Biológicas, Universidade Federal de Minas Gerais, Belo Horizonte, Brazil*

JÜRGEN BERNHAGEN • *Vascular Biology, Institute for Stroke and Dementia Research (ISD), Klinikum der Universitaet Muenchen (KUM), Ludwig-Maximilians-University (LMU), Munich, Germany; Munich Heart Alliance, Munich, Germany; Munich Cluster for Systems Neurology (SyNergy), Munich, Germany*

OMAR EL BOUNKARI • *Vascular Biology, Institute for Stroke and Dementia Research (ISD), Klinikum der Universitaet Muenchen (KUM), Ludwig-Maximilians-University (LMU), Munich, Germany*

RICHARD BUCALA • *Department of Internal Medicine, Yale School of Medicine, New Haven, CT, USA*

ALLYSSON CRAMER • *Departamento de Bioquímica e Imunologia, Instituto de Ciências Biológicas, Universidade Federal de Minas Gerais, Belo Horizonte, Brazil*

SARAH J. CREED • *Monash Micro Imaging, Hudson Institute of Medical Research, Clayton, VIC, Australia*

NADIA S. DEEN • *The Ritchie Centre, Hudson Institute of Medical Research, Clayton, VIC, Australia*

SANJA DESPOTOVIĆ • *Institute for Biological Research "Sinisa Stankovic", University of Belgrade, Beograd, Serbia*

KIRSTIN D. ELGASS • *Monash Micro Imaging, Hudson Institute of Medical Research, Clayton, VIC, Australia*

LAURA FARR • *University of Virginia School of Medicine, Charlottesville, VA, USA*

JACQUELINE K. FLYNN • *Rheumatology Group, Centre for Inflammatory Diseases, Department of Medicine, School of Clinical Sciences at Monash Health, Faculty of Medicine, Nursing and Health Sciences, Monash Medical Centre, Monash University, Clayton, VIC, Australia*

IZABELA GALVÃO • *Departamento de Bioquímica e Imunologia, Instituto de Ciências Biológicas, Universidade Federal de Minas Gerais, Belo Horizonte, Brazil*

AMANDA GRAHAM • *University of Kansas Medical Center, Kansas City, KS, USA*

KATRIN GRUNER • *Institute for Biology I, Unit of Plant Molecular Cell Biology, RWTH Aachen University, Aachen, Germany*

RAN GU • *Department of Biochemistry and Molecular Medicine, School of Medicine, University of California at Davis, Sacramento, CA, USA; Institute for Pediatric Regenerative Medicine, Shriners Hospitals for Children-Northern California, UC Davis School of Medicine, Sacramento, CA, USA*

JAMES HARRIS • *Rheumatology Group, Centre for Inflammatory Diseases, Department of Medicine, School of Clinical Sciences at Monash Health, Faculty of Medicine, Nursing and Health Sciences, Monash Medical Centre, Monash University, Clayton, VIC, Australia*

MICHAEL J. HICKEY • *Centre for Inflammatory Diseases, School of Clinical Sciences at Monash Health, Faculty of Medicine, Nursing and Health Sciences, Monash Medical Centre, Monash University, Clayton, VIC, Australia*

ADRIAN HOFFMANN • *Vascular Biology, Institute for Stroke and Dementia Research (ISD), Klinikum der Universitaet Muenchen (KUM), Ludwig-Maximilians-University (LMU), Munich, Germany; Department of Anaesthesiology, Klinikum der Universitaet Muenchen (KUM), Ludwig-Maximilians-University (LMU), Munich, Germany*

TALI LANG • *The Szalmuk Family Department of Medical Oncology, Level 2, Cabrini Institute, Malvern, VIC, Australia*

JACINTA P. LEE • *Department of Medicine, University of Cambridge, Cambridge, UK*

LIN LENG • *Department of Internal Medicine, Yale School of Medicine, New Haven, CT, USA*

SHANNON MOONAH • *Department of Medicine, University of Virginia School of Medicine, Charlottesville, VA, USA*

ERIC F. MORAND • *Rheumatology Group, Centre for Inflammatory Diseases, Department of Medicine, School of Clinical Sciences at Monash Health, Faculty of Medicine, Nursing and Health Sciences, Monash Medical Centre, Monash University, Clayton, VIC, Australia*

M. URSULA NORMAN • *Centre for Inflammatory Diseases, Department of Medicine, School of Clinical Sciences at Monash Health, Faculty of Medicine, Nursing and Health Sciences, Monash Medical Centre, Monash University, Clayton, VIC, Australia*

WARREN B. NOTHNICK • *University of Kansas Medical Center, Kansas City, KS, USA*

RALPH PANSTRUGA • *Institute for Biology I, Unit of Plant Molecular Cell Biology, RWTH Aachen University, Aachen, Germany*

ANITA A. PINAR • *Department of Pharmacology, Faculty of Medicine, Nursing and Health Sciences, Monash University, Clayton, VIC, Australia*

INA RUDLOFF • *The Ritchie Centre, Hudson Institute of Medical Research, Clayton, VIC, Australia; Department of Paediatrics, Monash University, Clayton, VIC, Australia*

TAMARA SAKSIDA • *Institute for Biological Research "Sinisa Stankovic", University of Belgrade, Beograd, Serbia*

DZMITRY SINITSKI • *Vascular Biology, Institute for Stroke and Dementia Research, Klinikum der Universitaet Muenchen, Ludwig-Maximilians-University (LMU) Munich, Munich, Germany*

EDWIN SIU • *Department of Internal Medicine, Yale School of Medicine, New Haven, CT, USA*

IVANA STOJANOVIĆ • *Faculty of Medicine, University of Belgrade, Institute of Histology and Embryology, Beograd, Serbia*

MARIA B. SUKKAR • *Graduate School of Health, Faculty of Health, University of Technology Sydney, Ultimo, NSW, Australia*

FABIEN B. VINCENT • *Rheumatology Group, Centre for Inflammatory Diseases, Department of Medicine, School of Clinical Sciences at Monash Health, Faculty of Medicine, Nursing and Health Sciences, Monash Medical Centre, Monash University, Clayton, VIC, Australia*

AITI VIZZINI • *Marine Immunobiology Laboratory, Department of Biological Chemical Pharmaceutical Science and Technology, University of Palermo, Palermo, Italy*

MILICA VUJIČIĆ • *Institute for Biological Research "Sinisa Stankovic", University of Belgrade, Beograd, Serbia*

KOJI WATANABE • *National Center for Global Health and Medicine, Shinjuku, Tokyo, Japan*

SHAHRZAD ZAMANI • *Centre for Innate Immunity and Infectious Diseases, Hudson Institute of Medical Research, Monash University, Clayton, VIC, Australia*

LEON CHRISTIAN ZWIßLER • *Vascular Biology, Institute for Stroke and Dementia Research (ISD), Klinikum der Universitaet Muenchen (KUM), Ludwig-Maximilians-University (LMU), Munich, Germany*

Chapter 1

Studying the Pro-Migratory Effects of MIF

Adrian Hoffmann, Leon Christian Zwißler, Omar El Bounkari, and Jürgen Bernhagen

Abstract

Macrophage migration inhibitory factor (MIF) is an upstream regulator of innate immunity and dysregulated MIF is a key mediator of acute and chronic inflammatory processes, autoimmune and cardiovascular diseases, as well as cancer. MIF is a pleiotropic cytokine with chemokine-like functions that has been designated as an atypical chemokine (ACK). It orchestrates leukocyte recruitment and migration into inflamed tissues through non-cognate interactions with the classical chemokine receptors CXCR2 and CXCR4, pathways that are further facilitated by MIF's cognate receptor CD74. Here, we describe two complementary methods that can be used to characterize immune cell migration and motility responses controlled by MIF and its receptors. These are the Transwell filter migration assay, also known as modified Boyden chamber assay, a two-dimensional (2D) device, and a matrix-based three-dimensional (3D) chemotaxis assay. The Transwell system is primarily suitable to study chemotactic cell transmigration responses toward a chemoattractant such as MIF through a porous filter membrane. The 3D chemotaxis setup enables for the cellular tracking of migration, invasion, and motility of single cells using live cell imaging.

Key words Macrophage migration inhibitory factor, MIF, Chemotaxis, Chemokinesis, Migration, Motility, Invasion, Transwell migration, 3D chemotaxis, Atypical chemokine, Boyden chamber, μ-Slide chemotaxis, Inflammation

1 Introduction

Chemotaxis is the directed migration of cells toward a chemoattractant gradient. It enables immune cells such as monocytes, lymphocytes, or dendritic cells to reach their target location in an inflamed tissue or at sites of injury or infection. Chemotactic migration differs from an undirected—random—migration response of a cell, a process also often referred to as chemokinesis [1].

More than 50 years ago a soluble lymphocyte-derived product was discovered that was later named macrophage migration-inhibitor factor (MIF) and found to inhibit the random migration

James Harris and Eric F. Morand (eds.), *Macrophage Migration Inhibitory Factor: Methods and Protocols*,
Methods in Molecular Biology, vol. 2080, https://doi.org/10.1007/978-1-4939-9936-1_1,
© Springer Science+Business Media, LLC, part of Springer Nature 2020

of guinea pig macrophages out of capillary tubes [2, 3]. As the migration-inhibitory capacity of these lymphocyte-derived supernatants was more recently re-interpreted as a desensitization effect of a chemotactic challenge, MIF is now considered the first chemokine to be discovered and is among the first (together with interleukin-1 and the type I interferons) cytokines to be studied. Successful cloning and purification of bioactive MIF [4, 5] led to extensive basic and clinical research during the following decades and has established MIF as an upstream regulator of innate immunity with pivotal roles in acute and chronic inflammation, autoimmune and cardiovascular diseases, as well as cancer [6–14].

Contrary to its eponymous name and despite lacking the conserved structural hallmarks of classical chemokines, MIF functions as a *bona fide* chemokine to modulate leukocyte recruitment and migration in inflamed tissues. This is achieved through non-cognate interaction with the classical chemokine receptors CXCR2, CXCR4, as well as the atypical chemokine receptor CXCR7/ACKR3. The interaction between MIF and its cognate receptor CD74 (MHC class II invariant chain, Ii) further modulates MIF-driven cell migratory responses in inflammatory diseases, but also serves to activate a variety of additional cell responses, including cell proliferation and cell metabolism that have been found to play a role in cancer, metabolic and ischemic heart disease [7, 15–17]. Despite the structural divergence, MIF has a functional similarity to classical chemokines and is classified as an atypical chemokine [18, 19]. From the point of discovery until today, the cell migratory/chemotactic potential of MIF on immune and other cell types has been of great interest due to its fundamental impact on a variety of immunologic processes and is still far from being elucidated.

We have established/optimized commonly used Transwell chemotaxis assays and 3D chemotaxis systems using μ-Slide chemotaxis chambers from ibidi GmbH to characterize the chemotactic capacity of MIF and its receptors toward various immune cell types, including primary monocytes, neutrophils, T cells, dendritic cells, and B cells, as well as cancer cell lines. In the framework of this methods article, we will focus on MIF-mediated migration responses of human monocytes and murine B cells.

The principle of the Transwell chemotaxis assay has originally been described by Boyden in 1962 [20]. It consists of two compartments, an upper and a lower chamber, that are separated by a cell-permeable porous membrane. Depending on the size of the studied cells, a variety of porous filters with the appropriate pore size can be used. Pore sizes are typically smaller than the diameters of the studied cells (e.g., 3 or 5 μm), promoting the detection of actively migrating versus randomly diffusing cells. The cells are applied on the upper side of the membrane, while the chemoattractant is added to the lower chamber. Cells that have migrated into the lower chamber following specific incubation times are counted and compared to control samples without chemoattractant ("the random or spontaneous migration control"). The Transwell migration assay is a sensitive method that can be used to investigate MIF-mediated immunological effects at low MIF concentrations in the low nanomolar range. The specific involvement of the MIF receptors can be easily examined by co-incubation with receptor-blocking antibodies or small molecule inhibitors of the chemokine receptors and/or MIF. Due to the relatively fast diffusion of the chemoattractant from the lower to the upper chamber, extended time-dependent studies are not suitable for this assay, if true chemotactic responses are to be studied.

The 3D chemotaxis assay complements the classical end-point Transwell assay as the cells of interest are embedded in a 3D gel matrix that can be observed via time-lapse microscopy after establishing a chemoattractant gradient. It therefore allows for measurements of early migration responses toward a maximum chemoattractant gradient and for speed measurements in the minute range. The ibidi μ-Slide 3D chambers mimic natural tissue-like cell environments, allow for a slower initialization and even maintenance of the chemoattractant gradient over time [21, 22]. Dose-dependent studies are more difficult to perform owing to the complex experimental setup and handling (see below). In addition, the analysis is less objective as only a relatively small number of cells can be tracked. Yet, the main advantage of the 3D assay is the possibility of tracking cell migration via live cell imaging not only to differentiate between chemotaxis and chemokinesis but also to quantify various chemotactic parameters [23]. A combination of the 2D Transwell and 3D μ-Slide assay creates a robust picture of MIF-induced pro-migratory effects and can be adapted for a variety of cell types.

2   Materials

We recommend using freshly prepared or aliquoted and non-expired solutions only. Unless indicated otherwise, store and handle solutions and materials as recommended by the manufacturer. Prepare solutions under sterile conditions using double-distilled water. Follow all waste disposal regulations, when disposing waste materials.

2.1   General

1. Cell culture incubator: 37 °C, 5% CO_2, high humidity.

2. Cell counting: Automated Cell Counter with appropriate counting slides from BioRad Laboratories and Trypan Blue Stain (0.4%). Other cell counting systems can otherwise be used.

3. Centrifuges.

4. Standard cell culture microscope.

5. Water bath, 37 °C.

6. Cell type-specific growth medium.

7. Slant tweezer.

8. Medium: Roswell Park Memorial Institute 1640 Medium (RPMI 1640) supplemented with 2 mM L-glutamine.

9. Treatment master mixes (MM): prepared in medium containing purified endotoxin-free, recombinant human or mouse MIF and MIF storage buffer as control (*see* **Note 1**). Optional: Additional chemoattractants can be used as a positive (or negative) control (e.g., CXCL8, CXCL12, CCL2, CXCL13).

10. Tubes: 1.5 mL, 15 mL and flow cytometry tubes.

2.2 Transwell Chemotaxis Assay

1. Transwell plates: 24 Well Cell Culture Cluster, Corning, REF 3526.

2. Transwell inserts: 6.5 mm Transwell with 5.0 μm Pore Polycarbonate Membrane Inserts.

3. Phosphate-buffered saline (PBS) without calcium and magnesium.

4. Counting beads: CountBright Absolute Counting Beads, Thermo Fisher Scientific, or other counting beads.

5. Flow cytometer with complementary tubes and solutions as described by the manufacturer.

2.3 3D Chemotaxis Assay

1. μ-Slide Chemotaxis, ibidi GmbH.

2. Petri dishes, 100 mm diameter, 20 mm height.

3. Gel MM: 10× Dulbecco's Modified Eagle's Medium (DMEM) (20 μL), NaOH 1 M (5 μL), H_2O (112 μL), $NaHCO_3$ 7.5% (3 μL), 1× DMEM (50 μL).

4. Collagen: collagen type 1, rat tail, ibidi GmbH (*see* **Note 2**).

5. 10–200 μL beveled pipet tips.

6. Fully motorized time-lapse inverted microscope (phase contrast) with additional heating system or stage top incubator, and optional autofocus.

7. Software: Chemotaxis and Migration Tool, Download Version 2.0.

8. Software: Image J (1.34k or later).

9. Manual tracking plugin (for Image J).

3 Methods

3.1 Transwell Chemotaxis

A schematic of the Transwell assay is shown in Fig. 1.

3.1.1 Sample Preparation

1. Cultivate the cells of choice, as usual (*see* **Note 3**).
2. Pre-warm medium in a water bath at 37 °C.
3. Label the lid of the Transwell plate according to the treatments planned (*see* **Note 4**).
4. Prepare treatment MM in pre-warmed medium. Calculate for 600 μL MM/well (*see* **Note 5**). After the preparation, place the medium in the water bath at 37 °C again.
5. Add 600 μL MM to the wells of the Transwell plate according to the experimental design (*see* **Note 6**).
6. Place one Transwell insert into each well by using a slant tweezer (*see* **Note 7**).
7. Close the lid and incubate the Transwell plate for 30 min in a cell culture incubator.
8. In the meantime, transfer the cells to an empty 15 mL tube, determine the cell number and cell viability (*see* **Note 8**).

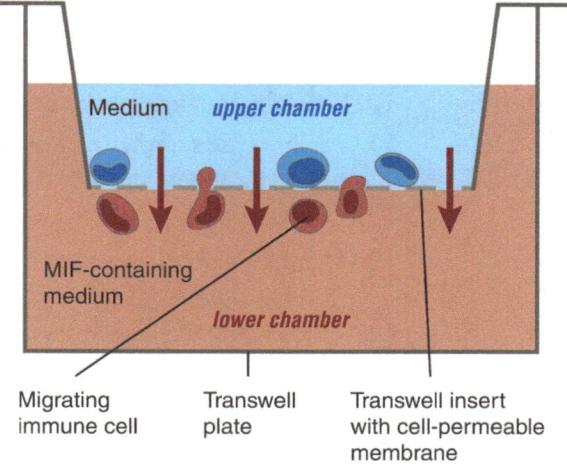

Fig. 1 Schematic drawing of the Transwell chemotaxis assay. Cells of interest (here depicted as monocytes) are seeded in chemoattractant-free medium in the upper chamber after MIF-containing medium has been added in the lower chamber of the Transwell plate. Upper and lower chamber are separated by a Transwell insert with a membrane that is permeable for the migrating cell type. Non- or slow-migrating cells as well as those just undergoing nondirectional spontaneous movements remain in the upper chamber. Depicted in blue are non-migrating cells or cells before the migration event. Depicted in red are cells that have migrated to the lower chamber

9. Centrifuge the cells at $300 \times g$ for 10 min, discard the supernatant and dilute the pellet in pre-warmed medium to a final concentration of 10^7 cells/mL.

10. Pipet 100 µL of the cell suspension (1×10^6 cells/well) produced in **step 9** into each Transwell insert without touching and damaging the membrane.

11. Place the Transwell plate for the desired migration time interval in the cell culture incubator (*see* **Note 9**).

3.1.2 Counting of Migrated Cells

1. Incubate the counting beads 30 min at RT prior to use.

2. Prepare the flow cytometer for the upcoming measurements.

3. After incubation (Subheading 3.1.1, **step 11**), carefully remove the Transwell inserts with a slant tweezer.

4. Optional: check the number and distribution of the cells in the well using a standard cell culture microscope (*see* **Note 10**).

5. Transfer the cell suspension from each well into labeled 1.5 mL tubes.

6. Wash each well with 500 µL PBS and add to the respective tube from **step 5**.

7. Centrifuge the tubes at $300 \times g$ for 5 min at room temperature (RT).

8. Discard 500 µL from the supernatant.

9. Add 25 µL of counting beads and mix gently by pipetting up and down.

10. Transfer the cells to the appropriate flow cytometer tubes.

11. Analyze the samples by flow cytometry using the same settings for each sample (*see* **Note 11**).

12. Gate populations of cells and beads.

13. Determine both number of cell events and number of bead events.

14. Calculate the cell number of the migrated cells for each sample using the following equation:

$$\text{Number of cells}/\mu L = \frac{\text{Number of cell events}}{\text{Number of bead events}} \times \frac{\text{Bead count of the lot } (\frac{\text{beads}}{25}\mu L)}{\text{Volume of sample } [\mu L]}.$$

3.2 3D Chemotaxis Assay

The method is essentially described by the manufacturer. Here we outline the handling and important steps for MIF-focused 3D chemotaxis assays using primary monocytes and B cells. This protocol uses the µ-Slide from ibidi (Fig. 2).

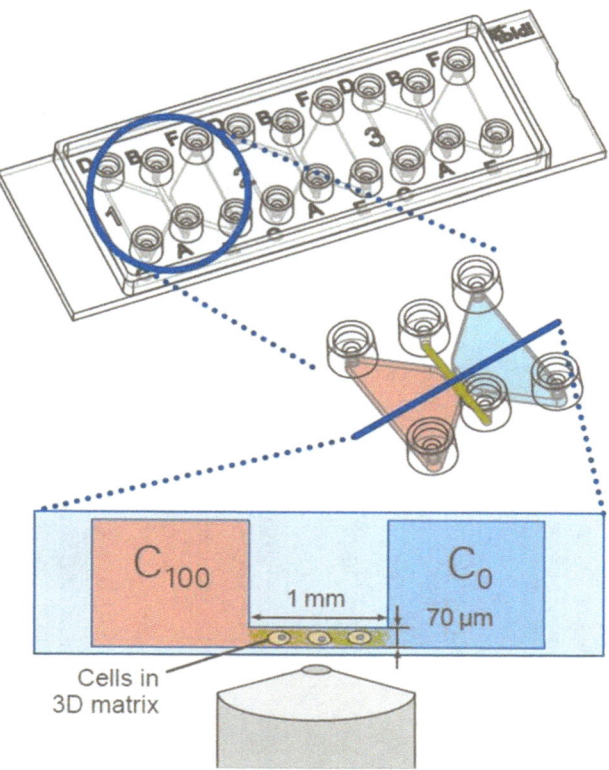

Fig. 2 Each μ-Slide consists of 3 identical chambers. The central channel of the μ-Slide chemotaxis chamber contains the cells of interest in a three-dimensional aqueous collagen-gel matrix. A stable chemoattractant gradient is established within the chamber through application of different chemoattractant concentrations (depicted as C_{100} to C_0) to the surrounding reservoirs. Migration of cells is visualized by time-lapse microscopy. (Figure reproduced by permission of ibidi GmbH)

3.2.1 Sample
Preparation

1. On the day before the experiment unpack the μ-Slide, place it into a Petri dish and close the filling ports C, D, E, F of the chemoattractant reservoirs (Fig. 3) using a slant tweezer and the provided plugs. Prepare a 15 mL tube containing medium (amount of medium depends on experimental setup). Place the Petri dish and the tube (with the cap loosely attached) in a cell culture incubator. Keep the μ-Slide placed in the Petri dish during the experiment and only take off the lid when needed (*see* **Note 12**).

2. Cultivate cells as appropriate.

3. Prepare and setup the microscope for the following measurements (*see* **Note 13**).

4. Place all collagen gel solutions (gel MM solutions and collagen) on ice for at least 10 min.

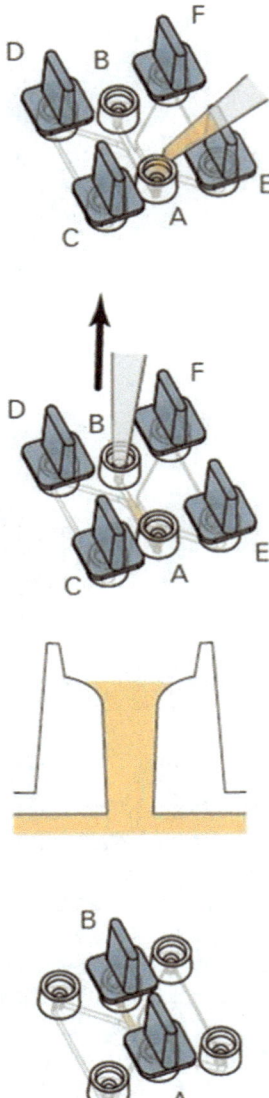

Fig. 3 Illustration of filling ports and essential pipetting steps handling the μ-Slide chemotaxis system. (Figure reproduced by permission of ibidi GmbH)

5. Determine cell number and cell viability. Prepare a cell suspension with a desired final concentration of 8×10^6 cells/100 μL medium (*see* **Note 14**).

6. Prepare the gel MM by adding all solutions in the order as listed in Subheading 2.3. Split the MM in 3 tubes of 63 μL each (*see* **Note 15**).

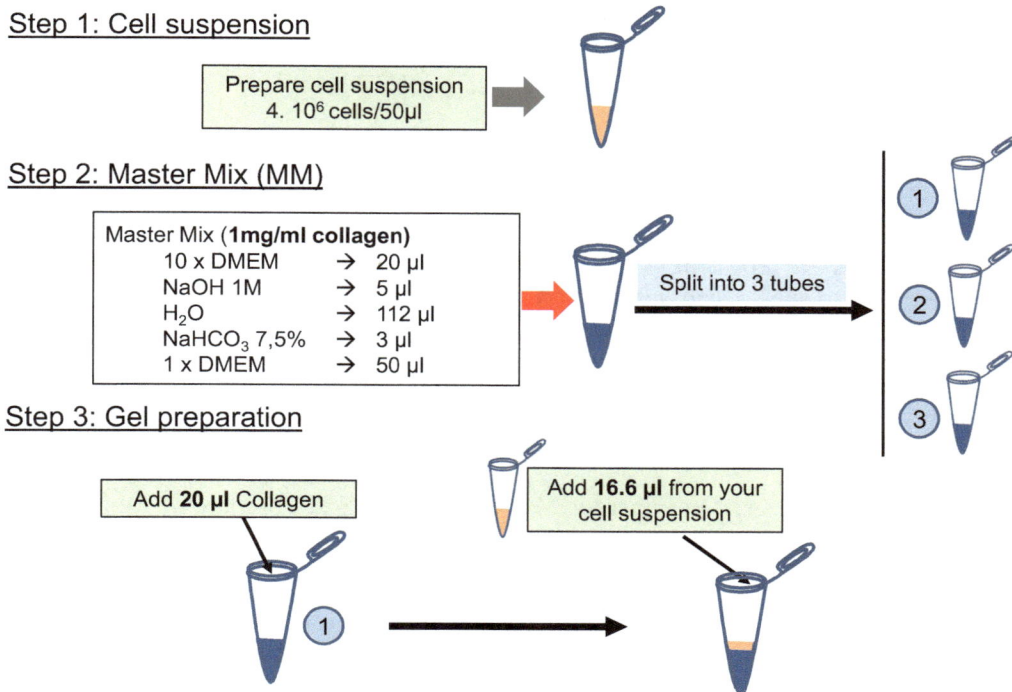

Step 1: Cell suspension

Prepare cell suspension
4. 10^6 cells/50µl

Step 2: Master Mix (MM)

Master Mix (**1mg/ml collagen**)
10 x DMEM	→	20 µl
NaOH 1M	→	5 µl
H_2O	→	112 µl
$NaHCO_3$ 7,5%	→	3 µl
1 x DMEM	→	50 µl

Split into 3 tubes

1
2
3

Step 3: Gel preparation

Add **20 µl** Collagen

1

Add **16.6 µl** from your cell suspension

Fig. 4 Pipetting scheme of cell suspension in collagen gel (step 4–10)

7. Place a small piece of wet paper tissue on one side of the Petri dish to create a humid atmosphere that reduces evaporation during gelation process.

8. Make sure that all filling ports except the middle channel ports A and B are closed with plugs (*see* Fig. 3).

9. Add 20 µL of collagen to one tube containing the gel MM. Resuspend thoroughly to create a homogeneous collagen-MM suspension (*see* **Note 16**).

10. Add 16.66 µL of the cell suspension from **step 5** to the collagen-gel MM suspension. Resuspend thoroughly to create a homogeneous mixture (*see* **Note 16**). A brief pipetting scheme of cell suspension in collagen gel (**steps 4–10**) is illustrated in Fig. 4 (*see* Fig. 4).

11. Pipet 6.3 µL of the collagen-gel MM-cell suspension from **step 10** to filling port A by angling the pipet about 45° and carefully dropping the mixture indirectly into the filling port (Fig. 3).

12. Aspirate by directly pressing the pipet tip into filling port B until the gel-cell mixture reaches the pipet tip (*see* **Note 17**) (Fig. 3).

13. Optional: Filling ports A and B should be evenly filled with the gel-cell mixture (Fig. 3). If this is not the case after performing

step 12, additional gentle aspiration steps from either filling port A or B should be included to adjust the filling levels.

14. Repeat steps 11, 12 (and 13) until all central channels are filled.

15. Gently remove the plugs from the chemoattractant reservoir ports C, D, E, F and close the central channel ports A, B (Fig. 3).

16. Optional: Check each channel for air bubbles and homogenous cell distribution using a standard cell culture microscope.

17. Incubate the slides in a cell culture incubator for 30–45 min.

18. Afterwards, ensure correct gel polymerization using a standard cell culture microscope. Only include intact channels without air bubbles for the experiment.

3.2.2 Addition of Chemoattractants, i.e., MIF

1. During step 17 of Subheading 3.2.1 prepare treatment MM in medium (see Note 18).

2. Close filling ports E and F of all chambers with plugs.

3. Carefully pipet 65 μL of medium (control medium, optional: medium containing chemoattractant storage buffer) to each filling port C by directly and vertically pressing the beveled pipet tip into the filling port until the respective reservoir is fully flushed with liquid. Optional: Slightly incline the μ-Slide so that filling port C faces downwards to avoid the formation of air bubbles.

4. Remove plugs from filling ports E and F, close filling ports C and D.

5. Carefully pipet 65 μL of medium to each filling port E by directly and vertically pressing the beveled pipet tip into the filling port until the respective reservoir is fully flushed with the liquid.

6. Add 15 μL of chemoattractant-containing medium to filling port E by angling the pipet about 45° and carefully dropping the medium indirectly into the filling port as described in step 11 of Subheading 3.2.1.

7. Aspirate 15 μL by directly pressing the beveled pipet tip into filling port F and discard the medium.

8. Repeat steps 6 and 7 once, so that a final volume of 30 μL chemoattractant-containing medium are applied to port E and 30 μL medium is discarded from port F (see Note 19).

9. Repeat steps 6–8 for the remaining chambers until all treatments are applied to the according reservoirs (see Note 20).

10. Close remaining ports E and F with plugs. Immediately continue with Subheading 3.2.3.

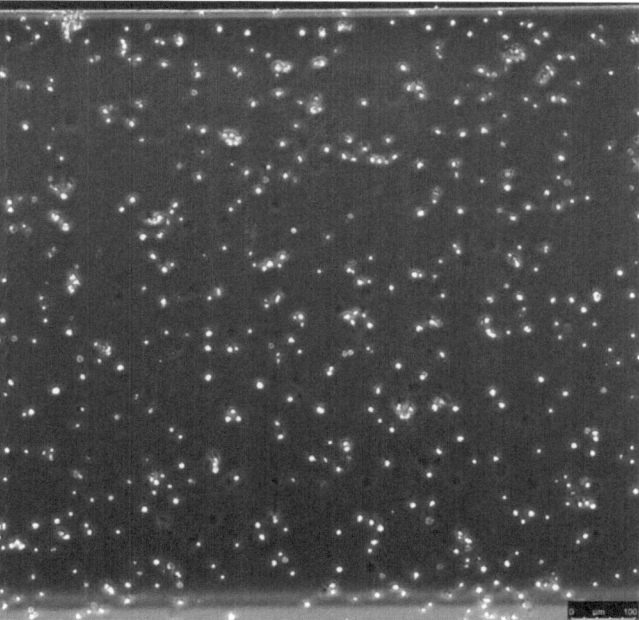

Fig. 5 Representative microscope image of a homogenous cell distribution of primary human monocytes within the central channel of the μ-Slide chemotaxis device

3.2.3 Microscopy

1. Install the chemotaxis slides on the motorized and pre-heated stage (optional: in a stage top incubator) of the microscope. Adjust the correct image positions of each channel as well as the focus and light intensity (*see* **Note 21**) (Fig. 5).

2. Perform time lapse microscopy for 2 h with a time interval of 1 min (*see* **Note 22**).

3. Save raw data and export files as uncompressed .avi or single image files.

3.2.4 Analysis

1. Import the uncompressed .avi file to Image J and open the Manual Tracking plug-in (*see* **Note 23**).

2. Print the first image of the video and randomly mark 40–50 cells (*see* **Note 24**).

3. Track each cell by selecting "add track" and follow the cell through all images taken by clicking on the cell. Track at least 30 cells (*see* **Note 25**).

4. Save the data table containing all tracks with the given x/y data sets by generating a tab-separated .xls file or .text (MS-DOS) file (*see* **Note 26**).

5. Import the data to the Chemotaxis and Migration Tool and click on "show original data."

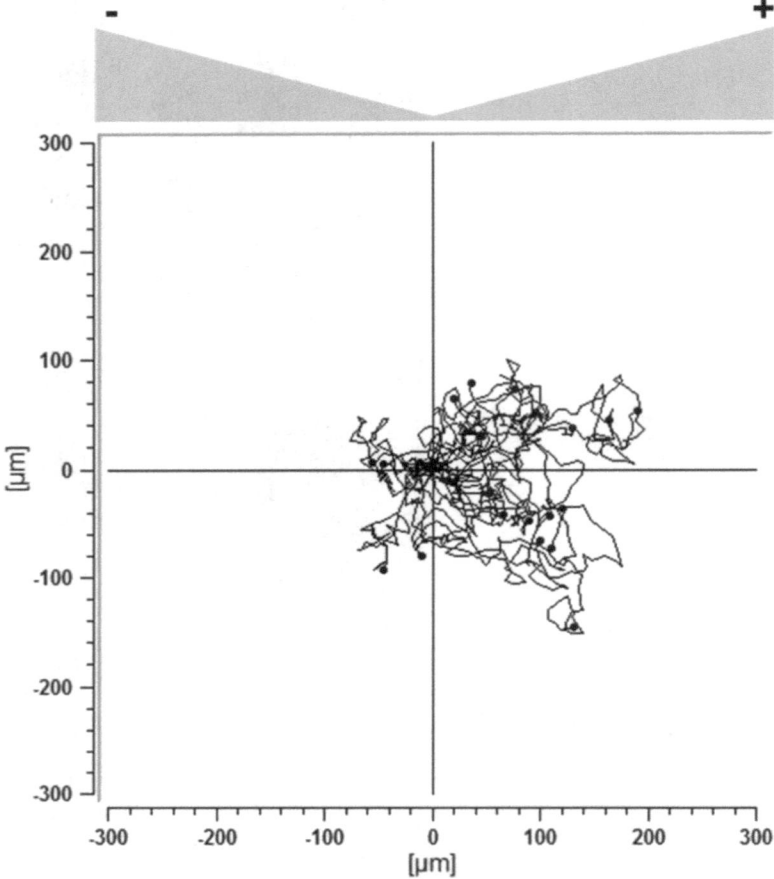

Fig. 6 Representative trajectory plot ($x, y = 0$ at time 0 h) of migrated monocytes toward a chemoattractant gradient with highest concentration of chemoattractant on the right side of the graph (depicted as +)

6. Initialize the data set by defining the number of slices (images taken per cell).

7. Calibrate x/y by defining the length of one pixel in μm and the time interval used.

8. Rotate the data sheet so that the control medium-containing side (−) is on the left side and the chemoattractant-containing side (+) on the right side of the graph (*see* **Note 27**) (Fig. 6).

9. Press "Apply settings" after the completed initialization.

10. Press "plot data" to create a graph.

11. The software includes a wide array of tools and possibilities to adjust graph settings and to obtain chemotactic parameters or statistics (i.e., forward migration index, center of mass, velocity), that can be exported and used for data presentation [16] (Fig. 6).

4   Notes

1. For the purification of bioactive human and mouse MIF we refer to published protocols [7, 24]. Using this purification method or slight variation of it, the purified recombinant MIF is considered to be endotoxin-free, as the endotoxin content is in the range of 5–15 pg LPS/µg rMIF (determined in the final enriched, sterile-filtered protein solution using the Pierce™ LAL Chromogenic Endotoxin Quantitation Kit). Purified recombinant MIF is resuspended in 20 mM sodium phosphate buffer, pH 7.2 (storage buffer). In general, purified refolded MIF should always be stored at 4 °C, not frozen and kept under sterile conditions. Due to precipitation that occurs in MIF stock solutions over time, the MIF protein concentration will decrease by approximately 5–20% per week of storage, which affects the bioactivity. Therefore, we always recommend to re-assess the concentration of the MIF stock solution on a regular basis. As a rule of thumb, refolded rMIF stock solutions should not be used for chemotaxis assays after 4–6 weeks of storage.

2. Our protocol was performed with collagen-1, rat tail from ibidi GmbH. Collagen from other manufacturers can be used, but gel composition may require optimization.

3. Cultivate your cells according to the cell type used. For information concerning the isolation, preparation, or cultivation of each cell type, we refer to method descriptions in state-of-the-art publications. In principal, the described migration assays can be performed with either primary cells or non-adherent immortalized cells/cell lines. In general, the migration efficiency in these assays is dependent on the cell type used, which may require protocol optimizations. Our protocol was optimized for MIF-induced migration of primary human monocytes and murine B lymphocytes. After primary cell isolation, we recommend to grow the cells overnight in a cell culture incubator (full growth medium, 37 °C, 5% CO_2) to avoid the internalization of chemokine receptors that otherwise may be induced by cell stress during the isolation/culturing process. The assays are performed in the designated cell culture medium but without the addition of fetal calf serum (FCS) or other growth factor-containing supplements. Starvation of cells may be required to increase the migratory responses of the cells, but it may affect the cell viability and is therefore not used in our protocol. In order to avoid contaminations, perform all steps under sterile conditions (i.e., cell culture conditions).

4. Strictly avoid any contamination of the Transwell plate such as by moving over the plate with hands or pipettes. Any random dirt can affect the results obtained in this assay in a variety of ways, for instance by damaging the membrane and thereby leading to more or less migration, respectively.

5. We recommend performing duplicates or triplicates of the treatments used. Always include negative controls that contain the storage buffer of the MIF stock solution, i.e., 20 mM sodium phosphate buffer, pH 7.2. For MIF dose-dependency experiments, we recommend to include physiologically or pathophysiologically realistic MIF-concentrations ranging from 1 nM to 80 nM (i.e., 1 nM, 2 nM, 4 nM, 8 nM, 16 nM, 32 nM, 80 nM). MIF, like other chemokines, typically induces bell-shaped chemotactic response curves due to desensitization effects. MIF-triggered chemotactic indices (CTX) for primary human monocytes and murine B cells are in the range of CTX = 1.7–3 for the optimal MIF concentration. The classical chemokine CXCL12 is regularly used as a positive control when studying monocytes and B cells that have abundant CXCR4 expression (CTX = 2.5–4) [7, 25].

6. This protocol refers to the experimental setup using 24 well Transwell plates and the according Transwell inserts. Manufacturers offer a wide range of plates and inserts differing in the number of wells, pore sizes, and additional materials. Necessary adjustments to the experimental setup regarding the cell number and seeding volume need to be taken into account. Of great impact is the correct pore size of the membrane, which is supposed to be smaller than the diameter of the cells to avoid unspecific "sneaking" of cells into the lower chamber.

7. Avoid any bubbles in the immediate environment of the inserts. Bubbles can affect the results of the assay, for instance by leading to less or no migration due to the reduced contact area between the cell suspension and the permeable membrane.

8. We recommend a cell viability of at least 80%, as a high number of stressed or dead cells may falsify the results.

9. The appropriate incubation time depends on the scientific question and experimental setup. We propose incubation times ranging from 2 to 4 h. Due to the fast equilibration of the chemoattractant concentration between the lower and upper chamber, extended time-dependent studies are not recommended.

10. This step often gives a quick impression of whether the experiment was methodically successful.

11. We recommend establishing a gating scheme for the according cell type and counting beads in advance. The correct

differentiation between cells and beads is facilitated by the uniform properties and size of the beads.

12. This step allows for proper degasing and equilibration to the incubator atmosphere and is aimed to reduce the risk of air bubbles during the experiment. When unexperienced with handling the assay, we recommend to prepare spare slides or channels as occurring air bubbles or an inhomogenous gel polymerization will result in the exclusion of this channel from the experiment.

13. If possible, pre-heat the sample area of the microscope to 37 °C or use a stage top incubator for optimal experimental conditions. We recommend using a 10× objective to obtain a sufficient resolution with enough cells per view. At this point we further recommend to pre-set the correct image positions (center of each slide-channel) and save a template for later experiments. This step reduces the time spent to adjust the correct positions and focus during the experiment, especially when observing several slides in parallel.

14. We recommend a cell viability of at least 80% as a high number of stressed or dead cells may falsify the results. For primary monocytes and B cells, we usually prepare a solution of 4×10^6 cells in 50 μL of medium. The assay requires highly concentrated cells, but only small volumes so that downsizing of the initial volume reduces the amount of cells needed. The aim is to obtain an even distribution of living cells in the gel matrix that still allows for appropriate cell discrimination during analysis. Therefore, the correct number of cells may need to be adjusted.

15. Splitting the MM in three separate tubes is included to generate spare MM in case of gel polymerization within the tube during the following pipetting procedure, unintentional creation of air bubbles or large-scale experiments that require more time (Fig. 4).

16. Make sure the collagen aliquot was placed on ice before to enable slow polymerization of the gel. In case the collagen-aliquot contains air bubbles shortly centrifuge the tube before handling. Once the collagen is added to the MM, the gel polymerization process starts. Gelation usually occurs within 5 min. A higher collagen concentration will result in thicker gels and quicker gel polymerization, which requires a faster handling. Even though this step is most time-sensitive and precautions have to be taken not to create any air bubbles, take enough time to thoroughly resuspend and stir the gel-cell suspension as it will result in a homogenous matrix. There exists a variety of possible gel compositions that are essentially described by the manufacturer and need to be adapted if another cell type is used. Following these steps is

essential to obtain a homogenous gel distribution within the channel and usually ensures that both filling ports are evenly filled. To avoid air bubbles in the channel, apply gentle pressure while aspirating. Only aspirate with the pipet during this step. Only use beveled pipet tips that allow a tight sealing of the filling port while aspirating.

17. When handling several slides at the same time we recommend to prepare a new collagen gel-cell mixture with the spare gel MM tubes every two slides. This reduces the risk of gelation during the pipetting process, but can be adapted when routine handling is ensured.

18. As the treatment MM will be diluted 1:2 during the application in **steps 5–10** of Subheading 3.2.2 *Addition of chemoattractants* (30 μL of chemoattractant-containing treatment MM + 30 μL of medium), prepare treatment MM with a concentration that is double of the final concentration. Calculate for a volume of at least 50 μL/treatment as 30 μL of chemoattractant-containing medium will be applied. Always include negative controls $(-/-)$ that contain the storage buffer of the MIF stock solution, respectively 20 mM sodium phosphate buffer. Consider including a control with the chemoattractant in both reservoirs $(+/+)$ for investigating its influence on chemokinesis (When setting up a $(+/+)$ chamber the principle from Subheading 3.2.2, **steps 6–8** have to be performed equally for the reservoir with the filling ports C and D). We recommend using MIF-concentrations in a physiologically or pathophysiologically realistic range between 1 nM and 80 nM depending on the scientific question. The classical chemokine CXCL12 is regularly used as a positive control in MIF-focused migration assays using monocytes and B cells with abundant CXCR4 expression.

19. This way of applying the chemoattractant-containing medium is aimed to a slower initialization of the chemoattractant gradient compared to the direct application into an empty reservoir. In case of early and fast migration of the cells this can help to save some important time.

20. When handling several slides, we recommend to prepare an experimental layout template to indicate the different treatments and to number the slides. This would help to indicate channels that cannot be used due to an inhomogenous gel or air bubbles.

21. Find an appropriate image position for each channel with an even distribution of cells and use this position for the time lapse microscopy (Fig. 5).

22. For fast migrating monocytes and B cells, we use a time-lapse stage experiment with a duration of 1–3 h and an image time

interval every 1 min. The time-lapse measurement can be adjusted to the specific aim of the experiment and to the cell type.

23. We describe a tracking method using the commonly used Manual tracking plugin from Image J. Other commercially available manual or automated tracking tools may be used.

24. This step is included as manual tracking will always lack a certain degree of objectivity. Additionally, we recommend blinding the examiner so that he does not know which treatment was used.

25. Exclude dying cells, cells that are not moving or leave the field of view by deleting their tracks.

26. Image J currently only allows to save the table as a .csv-file. In case of problems importing the data set to the Chemotaxis and Migration Tool, we recommend to copy the raw data set manually into an excel file and then export as .text (MS-DOS) file (columns from left to right should be formatted as following: total number of slices starting from 1 to the exact final number, track number, slice number of the respective track, x, y). This step is necessary as not all formats are accepted by the Chemotaxis and Migration Tool.

27. To identify the correct treatment sides of the image, remember that the image is inverted.

Acknowledgements

This work was supported by the Metiphys scholarship of the Ludwig-Maximilians-University (LMU) Munich to A.H. and by Deutsche Forschungsgemeinschaft (DFG) grants SFB1123-A03, BE 1977/10-1, and BE 1977/11-1 to J.B. It was co-supported by the Wilhelm-Sander-Stiftung grant 2017.009.1 and the Else-Kröner-Fresenius-Stiftung (EKFS) grant 2014/A216 to J.B. as well as by DFG under Germany's Excellence Strategy within the framework of the Munich Cluster for Systems Neurology (EXC 2145 SyNergy—ID 390857198) to J.B. We thank ibidi GmbH (Martinsried, Germany) for providing us with the original figures illustrating the 3D μ-Slide chemotaxis assay.

References

1. Bourne HR, Weiner O (2002) A chemical compass. Nature 419(6902):21
2. David JR (1966) Delayed hypersensitivity in vitro: its mediation by cell-free substances formed by lymphoid cell-antigen interaction. Proc Natl Acad Sci U S A 56(1):72–77
3. Bloom B, Bennett B (1966) Mechanism of a reaction in vitro associated with delayed-type hypersensitivity. Science 153:80–82
4. Bernhagen J et al (1993) MIF is a pituitary-derived cytokine that potentiates lethal endotoxaemia. Nature 365(6448):756–759

5. Weiser WY et al (1989) Molecular cloning of cDNA encoding a human macrophage migration inhibtion factor. Proc Natl Acad Sci U S A 86:7522–7526

6. Bernhagen J et al (1996) An essential role for macrophage migration inhibitory factor in the tuberculin delayed-type hypersensitivity reaction. J Exp Med 183(1):277–282

7. Bernhagen J et al (2007) MIF is a noncognate ligand of CXC chemokine receptors in inflammatory and atherogenic cell recruitment. Nat Med 13(5):587–596

8. Calandra T et al (1995) MIF as a glucocorticoid-induced modulator of cytokine production. Nature 377(6544):68–71

9. Calandra T et al (2000) Protection from septic shock by neutralization of macrophage migration inhibitory factor. Nat Med 6(2):164–170

10. Calandra T, Roger T (2003) Macrophage migration inhibitory factor: a regulator of innate immunity. Nat Rev Immunol 3(10):791–800

11. Donnelly SC et al (1997) Regulatory role for macrophage migration inhibitory factor in acute respiratory distress syndrome. Nat Med 3:320–323

12. Morand EF, Leech M, Bernhagen J (2006) MIF: a new cytokine link between rheumatoid arthritis and atherosclerosis. Nat Rev Drug Discov 5(5):399–410

13. Tilstam PV et al (2017) MIF family cytokines in cardiovascular diseases and prospects for precision-based therapeutics. Expert Opin Ther Targets 21(7):671–683

14. Zernecke A, Bernhagen J, Weber C (2008) Macrophage migration inhibitory factor in cardiovascular disease. Circulation 117(12):1594–1602

15. Leng L et al (2003) MIF signal transduction initiated by binding to CD74. J Exp Med 197(11):1467–1476

16. Alampour-Rajabi S et al (2015) MIF interacts with CXCR7 to promote receptor internalization, ERK1/2 and ZAP-70 signaling, and lymphocyte chemotaxis. FASEB J 29(11):4497–4511

17. Sinitski D et al (2019) Macrophage migration inhibitory factor (MIF)-based therapeutic concepts in atherosclerosis and inflammation. Thromb Haemost 119(4):553–566

18. Kapurniotu A, Gokce O, Bernhagen J (2019) The multitasking potential of alarmins and atypical chemokines. Front Med (Lausanne) 6:3

19. Tillmann S, Bernhagen J, Noels H (2013) Arrest functions of the MIF ligand/receptor axes in atherogenesis. Front Immunol 4:115

20. Boyden S (1962) The chemotactic effect of mixtures of antibody and antigen on polymorphonuclear leucocytes. J Exp Med 115:453–466

21. Biswenger V et al (2018) Characterization of EGF-guided MDA-MB-231 cell chemotaxis in vitro using a physiological and highly sensitive assay system. PLoS One 13(9):e0203040

22. Zengel P et al (2011) mu-slide chemotaxis: a new chamber for long-term chemotaxis studies. BMC Cell Biol 12:21

23. Pepperell EE, Watt SM (2013) A novel application for a 3-dimensional timelapse assay that distinguishes chemotactic from chemokinetic responses of hematopoietic CD133(+) stem/progenitor cells. Stem Cell Res 11(2):707–720

24. Bernhagen J et al (1994) Purification, bioactivity, and secondary structure analysis of mouse and human macrophage migration Inhibitory factor (MIF). Biochemistry 33:14144–14155

25. Klasen C et al (2014) MIF promotes B cell chemotaxis through the receptors CXCR4 and CD74 and ZAP-70 signaling. J Immunol 192(11):5273–5284

Chapter 2

In Vitro Approaches for Investigating the Influence of MIF on Leukocyte-Endothelial Cell Interactions Under Flow Conditions

Michael J. Hickey

Abstract

The exit of leukocytes from the bloodstream into inflamed sites involves a sequence of interactions with vascular endothelial cells, in which leukocytes, moving rapidly in flowing blood, first tether and roll on the endothelial surface before arresting and then transmigrating across the endothelial barrier. Examining the mechanisms of these interactions in human systems has involved the use of in vitro flow chamber assays, using a variety of cells and immobilized molecules as adhesive substrata. Here we describe how to perform these assays using human umbilical vein endothelial cells and human leukocytes.

Key words Leukocyte interactions, Rolling, Adhesion, Transmigration, Flow chamber, Shear conditions

1 Introduction

A growing body of evidence now indicates that macrophage migration inhibitory factor (MIF) promotes leukocyte recruitment during inflammatory responses. Much of this evidence has been derived in vivo in mouse models in which MIF is absent (MIF-deficient mice) or inhibited [1–4]. Alternatively, local administration of exogenous MIF has been shown to promote leukocyte recruitment [5, 6] (*see* Chapter 3). However in vivo studies of this nature provide minimal information as to the cell type-specific effects of MIF which underlie this effect.

In order to address the endothelial cell-specific effects of MIF, we previously performed experiments with human umbilical endothelial cells (HUVEC) in which we knocked down expression of MIF using siRNA, and examined the capacity of these endothelial cells to support interactions with leukocytes under flow conditions [7]. These experiments revealed that MIF promotes endothelial cell

James Harris and Eric F. Morand (eds.), *Macrophage Migration Inhibitory Factor: Methods and Protocols*,
Methods in Molecular Biology, vol. 2080, https://doi.org/10.1007/978-1-4939-9936-1_2,
© Springer Science+Business Media, LLC, part of Springer Nature 2020

adhesive function via cell-intrinsic effects on adhesion molecule and chemokine expression.

In this chapter, we describe the use of HUVEC in flow chamber experiments for assessment of interactions with human leukocytes under flow conditions.

2 Materials

2.1 HUVEC Culture Reagents

1. Complete medium: M199 medium, fetal calf serum (20%), L-glutamine (2 mM), penicillin/streptomycin (1 U/mL and 0.1 mg/mL, respectively), endothelium mitogen (extract of bovine hypothalamus—50 μg/mL), heparin (0.1 mg/mL) (*see* **Note 1**).

2. Collagen (0.2% in PBS).

3. Fibronectin (0.025 mg/mL in PBS), collagenase II (0.1% in HBSS).

4. Human TNF (0.03–1.0 ng/mL final concentration) or IL-1β (10–100 U/mL final concentration).

5. Hanks Balanced Salt Solution (HBSS).

6. Phosphate Buffered Saline (PBS) (comprises NaCl—8.77 mg/mL, Na_2HPO_4 anhydrous—1.23 mg/mL, NaH_2-$PO_4 \cdot 2H_2O$—0.187 mg/mL; pH 7.4).

2.2 Human Leukocytes

A range of preparations can be used:

1. Whole blood, typically diluted 1:10 in HBSS [7–9].

2. Peripheral blood-derived mononuclear cells (PBMCs) [10].

3. Isolated neutrophils [11], or any other isolated leukocyte population of interest (*see* **Note 2**).

2.3 Culture Dishes

1. For experiments involving the Glycotech® flow chamber, it is recommended to use 35 mm dishes from Corning Life Sciences.

2.4 Parallel Flow Plate Chamber

1. Flow chamber. Here we describe use of the Glycotech® (Gaithersburg, MA, USA) circular flow chamber system designed for use in 35 mm dishes.

2. Silicon rubber gasket (flow width = 0.50 cm, thickness = 0.010 in).

3. Silastic™ laboratory tubing (1/16 in. ID).

4. Silicon vacuum grease.

2.5 Other Equipment

1. Syringe pump.

2. Vacuum pump.

2.6 Microscope	1. Inverted live cell imaging microscope and compatible video rate image capture system.

3 Methods

3.1 HUVEC Isolation and Culture

1. Pre-coat culture flasks with collagen (0.2% in PBS). Incubate collagen solution in plates for 30–60 min at room temperature. Discard collagen solution prior to adding cells.

2. Isolate HUVEC from umbilical cords as previously described [12] (*see* **Note 3**). In brief, insert a large bore needle into the umbilical vein, and flush the vessel with 3×20 mL warmed PBS. Clamp the lower end of the vein and fill it with collagenase II solution. Incubate the umbilical vein at 37 °C for 20 min. Drain out the collagenase and then harvest the HUVEC from the vein via perfusion with FCS-containing medium. Spin the endothelial cells down from the medium (1100 rpm/~ $250 \times g$, 9 min), then resuspend them in media and plate them out.

3. Culture isolated HUVEC in 0.2% collagen-coated culture flasks with complete medium. Cells from a single cord are typically seeded in a T25 flask to facilitate cell-cell contact between neighboring cells and generation of a monolayer (*see* **Notes 4** and **5**).

3.2 Flow Chamber Experiments

Experiments referred to herein describe the use of the circular Glycotech® flow chamber system that we have used for analysis of the role of MIF in endothelial cell biology (*see* **Note 6**). This system entails culture of endothelial cells in 35 mm dishes pre-coated with fibronectin.

1. On the day prior to flow chamber experiments, pre-coat 35 mm dishes with fibronectin (30–60 min at room temperature). Carefully remove fibronectin solution with a pipette tip and plate HUVEC into dishes at high density (*see* **Note 7**).

2. 3–4 h prior to use in flow chambers, activate HUVEC monolayers via exposure to inflammatory mediators such as TNF or IL-1β (*see* **Note 8**).

3. Calculate the syringe pump flow rate required to achieve the shear rate required for experiments, taking account of the size of the syringe being used in the pump (*see* **Note 9**). Program the flow rate into the syringe pump.

4. Apply the appropriate silicon gasket to the basal surface of the circular flow chamber, keeping it in place with a small amount of silicon grease. Connect the chamber to a vacuum pump via tubing.

5. Prepare a 35 mm dish containing a HUVEC monolayer for application of the flow chamber by filling it with 2–3 mL of HBSS. With the vacuum running, mount the chamber into the 35 mm dish. When loaded correctly, the vacuum should seal the chamber onto the 35 mm dish and the endothelial cell viewing chamber should be free of bubbles.

6. Place the chamber on to the stage of an inverted phase contrast microscope for viewing, ideally within a 37 °C live cell imaging chamber. Visualize and focus on the endothelial monolayer using the microscope.

7. Connect inflow and outflow tubing to the chamber, with the inflow connected to the Falcon tube containing the cell suspension (preferably maintained at 37 °C) and the outflow tubing connected to the needle/syringe in the syringe pump.

8. Using the syringe pump "Refill" command, draw leukocyte preparation into the chamber and across the endothelial surface at the defined flow/shear rate (*see* **Note 9**). Record images of interacting cells (*see* **Notes 10** and **11**).

3.3 Data Analysis

1. Leukocyte rolling. Count the number of cells rolling in a given field, i.e., undergoing a rotational interaction with the endothelial surface in the direction of shear flow. Express the data as cells/mm^2.

2. Leukocyte adhesion. Determine the number of leukocytes that remain static on the endothelial surface for the duration of the 10 s recording. If the recording is longer, 30 s can be used as a duration of attachment to be defined as adhesion. Express the data as cells/mm^2.

3. Leukocyte transmigration. If doing time-lapse imaging to study transmigration, cells can be visualized as they undergo transmigration. Alternatively, in experiments where brief recordings are made of multiple individual fields of view, transmigrated cells can be identified by their phase dark appearance, and often by their migration underneath the monolayer. In both cases, transmigration is expressed as cells/mm^2.

4   Notes

1. As an alternative to complete medium, endothelial growth medium (Lonza, Australia) can be used.

2. If performing experiments under flow conditions, leukocytes must retain the capacity to tether and roll on the endothelial surface, for subsequent adhesion and transmigration events to be assessed. Lack of care during leukocyte isolation procedures can lead to some cell populations losing the necessary

molecules for rolling (e.g., L-selectin, PSGL-1) from their surface. This will reduce the number of cells able to interact with the endothelial monolayer under flow conditions, limiting the utility of the experiment.

3. For more detailed information on how to isolate, culture, and utilize HUVEC, see the following publications [12, 13]. For use in in vitro flow-based leukocyte recruitment assays, primary or early passage HUVEC provide the best results. Later-passage HUVEC have reduced capacity to support leukocyte interactions under flow conditions [14].

4. We and others have shown that primary HUVEC are amenable to gene knockdown procedures such as siRNA as well as trans-genic approaches for induction of gene expression [7, 15]. Techniques such as electroporation can be used for introduction of genetic constructs into these cells. HUVEC retain functional capacity following these procedures, as demonstrated by their capacity to respond to inflammatory activation and induce leukocyte capture, adhesion, and trans-migration. Further information on these gene modulation approaches can be found in these publications [7, 15]. We used this approach previously to downregulate expression of MIF in HUVEC in order to demonstrate its endothelial cell-intrinsic functions.

5. Additional useful approaches for assessment of endothelial cell biology in the context of leukocyte recruitment include flow cytometry for assessment of surface adhesion molecule expression [7], and stable NF-κB reporter human microvascular endothelial cell (HMEC) cell lines for assessment of the mechanisms of endothelial cell activation [15].

6. Numerous devices are now available for performing flow chamber assays. A great deal of work has been done using parallel plate flow chambers and more straightforward flow chambers such as the Glycotech system, which fit into commercially available culture dishes. However, microfluidic devices are now favored for this type of work due to their relative ease of use and the fact that they require lower numbers of cells and smaller amounts of reagents.

7. Typically HUVEC are used at $\sim 2 \times 10^5/100$ μL ($2 \times 10^6/$ mL) of medium. To reduce the number of cells required per experiment and increase the number of experiments able to be performed from a single HUVEC preparation, cells can be plated into a small region in the middle of the 35 mm dish, rather than the entire dish surface. This region should be defined by the area covered during the fibronectin coating step. Its dimensions should be slightly larger than that of the rectangular viewing window of the flow chamber.

8. HUVEC activation is essential to induce robust leukocyte-endothelial cell interactions. This is typically achieved using inflammatory cytokines such as TNF or IL-1β. The level of activation, and therefore leukocyte-endothelial cell interaction, is concentration-dependent. We have used TNF in the range of 0.03–1.0 ng/mL [15]. IL-1β has been used at 10–100 U/mL [16, 17].

9. Ideally, experiments in flow chamber systems utilize shear rates that mimic those that occur in vivo in vessels where leukocyte-endothelial cell interactions occur, i.e., postcapillary venules. These interactions are highly shear-dependent, in that at high levels of shear, interactions will not occur while at shear levels below those that occur in vivo, interactions will readily occur but these interactions may not be physiologically relevant. In our experiments, we have predominantly performed experiments at $150\ s^{-1}$, a rate at which (1) TNF-activated HUVEC, but not resting HUVEC, readily support leukocyte interactions and (2) the classical three-step sequence of interactions (rolling, adhesion, transmigration) can be observed [7, 15].

10. In experiments using diluted whole blood, the blood is drawn into the chamber for a sufficiently long time to deliver leukocytes into the field of view and initiate a reasonable number of leukocyte interactions. However, as the erythrocytes obscure the view of the leukocytes, after this period the infusion is switched to straight HBSS. This serves to clear the erythrocytes from the chamber. Once the chamber is clear and the interacting leukocytes are readily visible, the recordings can be commenced.

11. For analysis of rolling and adhesion interactions, video-rate recordings are optimal. For analysis of adhesion, recordings up to 10 randomly selected fields for 10 s each allows a reasonably comprehensive sampling of the monolayer. For analysis of leukocyte transmigration, time-lapse recordings over longer durations (up to 10 min) will facilitate detection and analysis of transmigration. Alternatively, transmigration can be assessed after a 10 min period, identifying transmigrated leukocytes by their characteristic phase-dark appearance [15].

Acknowledgments

This work is supported by funding from the National Health and Medical Research Council of Australia (1042775, MJH) and the Monash University School of Clinical Sciences at Monash Health (MUN).

References

1. Gregory JL et al (2004) Reduced leukocyte-endothelial cell interactions in the inflamed microcirculation of macrophage migration inhibitory factor-deficient mice. Arthritis Rheum 50:3023–3034

2. Fan H et al (2011) Macrophage migration inhibitory factor and CD74 regulate macrophage chemotactic responses via MAPK and Rho GTPase. J Immunol 186(8):4915–4924

3. Gregory JL et al (2009) Independent roles of macrophage migration inhibitory factor and endogenous, but not exogenous glucocorticoids in regulating leukocyte trafficking. Microcirculation 16(8):735–748

4. Stark K et al (2013) Capillary and arteriolar pericytes attract innate leukocytes exiting through venules and 'instruct' them with pattern-recognition and motility programs. Nat Immunol 14(1):41–51

5. Gregory JL et al (2006) Macrophage migration inhibitory factor induces macrophage recruitment via CC chemokine ligand 2. J Immunol 177(11):8072–8079

6. Bernhagen J et al (2007) MIF is a noncognate ligand of CXC chemokine receptors in inflammatory and atherogenic cell recruitment. Nat Med 13(5):587–596

7. Cheng Q et al (2010) Macrophage migration inhibitory factor increases leukocyte-endothelial interactions in human endothelial cells via promotion of expression of adhesion molecules. J Immunol 185(2):1238–1247

8. Reinhardt PH, Kubes P (1998) Differential leukocyte recruitment from whole blood via endothelial adhesion molecules under shear conditions. Blood 92(12):4691–4699

9. Cheng Q et al (2012) Lymphocytes from systemic lupus erythematosus patients display increased spreading on VCAM-1, an effect associated with active renal involvement. Lupus 21(6):632–641

10. Lockmann A, Schon MP (2013) Phenotypic and functional traits of peripheral blood mononuclear cells retained by controlled cryopreservation: implications for reliable sequential studies of dynamic interactions with endothelial cells. Exp Dermatol 22(5):358–359

11. Reinhardt PH, Elliott JF, Kubes P (1997) Neutrophils can adhere via alpha4beta1-integrin under flow conditions. Blood 89:3837–3846

12. Woodman RC et al (1993) Effects of human neutrophil elastase (HNE) on neutrophil function in vitro and in inflamed microvessels. Blood 82:2188–2195

13. Parsons SA et al (2013) Studying leukocyte recruitment under flow conditions. Methods Mol Biol 946:285–300

14. Kanwar S et al (1995) Desmopressin induces endothelial P-selectin expression and leukocyte rolling in post-capillary venules. Blood 86:2760–2766

15. Cheng Q et al (2013) GILZ overexpression inhibits endothelial cell adhesive function through regulation of NF-kappaB and MAPK activity. J Immunol 191(1):424–433

16. Abe Y, Ballantyne CM, Smith CW (1996) Functions of domain 1 and 4 of vascular cell adhesion molecule-1 in alpha4 integrin-dependent adhesion under static and flow conditions are differentially regulated. J Immunol 157(11):5061–5069

17. Morigi M et al (1995) Fluid shear stress modulates surface expression of adhesion molecules by endothelial cells. Blood 85(7):1696–1703

Chapter 3

Using Intravital Microscopy to Study the Role of MIF in Leukocyte Trafficking In Vivo

M. Ursula Norman and Michael J. Hickey

Abstract

In vivo visualization of the microvasculature of the mouse cremaster muscle has been fruitful in the evaluation of the role of macrophage migration inhibitory factor in promotion of leukocyte trafficking. Here we explain how to undertake this preparation, including details on mouse anesthesia, securing intravenous access, and cremaster muscle exteriorization. We also provide information on the various microscopy modalities now available for imaging microvascular preparations of this nature.

Key words Intravital microscopy, Leukocyte interactions, Rolling, Adhesion, Transmigration, Cremaster muscle, Postcapillary venule

1 Introduction

Macrophage migration inhibitory factor (MIF) is a pleiotropic protein with numerous roles in the inflammatory response [1, 2]. This is evidenced by experiments in which inhibition or absence of MIF results in attenuation of inflammatory responses [3–6]. Studies from our laboratory and others have demonstrated that one aspect of inflammation controlled by MIF is leukocyte recruitment—a hallmark feature of inflammation [7–10]. In most inflammatory or infectious responses, leukocytes enter the affected site from the bloodstream, using variations of the classical rolling/adhesion/transmigration paradigm to interact with the endothelial surface of the microvasculature at the affected site [11, 12]. In a series of studies, we demonstrated a role for MIF in promotion of recruitment of neutrophils and monocyte-lineage cells from the bloodstream, via a wide range of mechanisms [7–9, 13–15]. In vivo, these experiments made use of intravital microscopy of the mouse cremaster muscle for investigation of leukocyte-endothelial cell interactions [7–9]. The observations from these in vivo experiments were supported by in vitro experiments examining the role of

James Harris and Eric F. Morand (eds.), *Macrophage Migration Inhibitory Factor: Methods and Protocols*,
Methods in Molecular Biology, vol. 2080, https://doi.org/10.1007/978-1-4939-9936-1_3,
© Springer Science+Business Media, LLC, part of Springer Nature 2020

cell-intrinsic MIF in endothelial cell function and leukocyte migration [13–15] (*see* Chapter 2).

In this chapter, we describe the use of intravital microscopy of the mouse cremaster muscle for assessment of the role of MIF in leukocyte-endothelial cell interactions in vivo.

2   Materials

2.1   Buffer

1. Bicarbonate superfusion buffer; 131.9 mM NaCl, 4.7 mM KCl, 1.2 mM $MgSO_4$, 20 mM $NaHCO_3$ in MilliQ H_2O. On the day of use, pre-warm the solution in the water bath (37 °C), then pH to 7.4 and keep solution in the water bath for the duration of the experiment (*see* **Note 1**).

2.2   Equipment

1. Peristaltic pump for superfusion buffer, e.g., Gilson Minipuls 3.

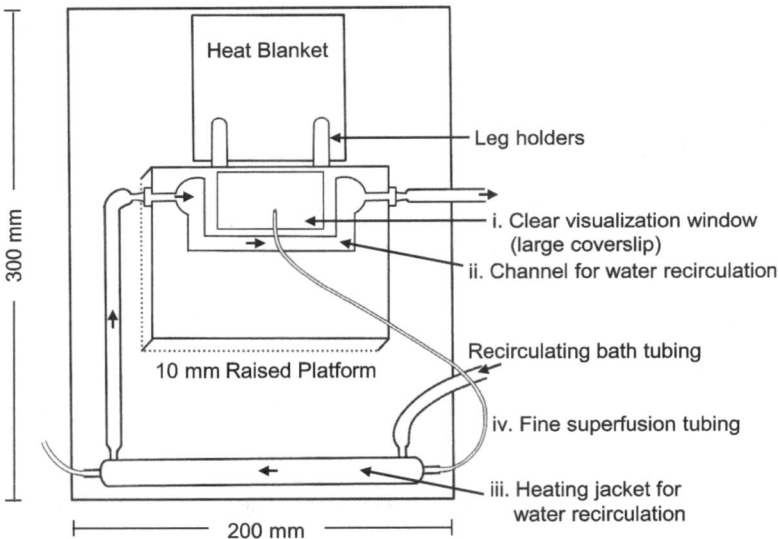

Fig. 1 Design of mouse cremaster platform for imaging via intravital microscopy. Diagram shows layout of mouse cremaster muscle board, constructed on a ~300 × 200 mm piece of plexiglas. The anesthetized mouse is placed on a heating blanket. Two plexiglas leg holders are used to secure the mouse in an appropriate position. The board incorporates a raised (~10 mm) platform which comprises (1) a clear visualization window (covered by a large coverslip) and (2) an enclosed channel for warm water recirculation. The circuit for water recirculation includes a tubular metal heating jacket (3), through which super-fusion fluid passes before reaching the muscle preparation. Superfusion fluid is pumped through the heating jacket and into fine superfusion tubing (4) before delivery to the muscle preparation

2. Circulating heated water bath (for heating the mouse platform), e.g., ThermoHaake C10-P5 Circulating Bath.

3. Imaging platform with transparent window (incorporating a cover slip) for transillumination of the cremaster muscle, e.g., our board is made from Plexiglas that has a hollowed-out section to run warm water from the circulating water bath to maintain the cremaster muscle in a warm environment (Fig. 1).

4. Heat mat to maintain the mouse's core body temperature throughout the experiment.

5. Intravital microscope (*see* **Note 2**).

2.3 Catheters

1. PE10 polyethylene tubing, mounted onto a 30G needle, pre-filled with saline from a 1.0 mL syringe.

2.4 Surgical Material and Instruments

1. Suture material; 4-0 silk.

2. Fine scissors, e.g., Lawton Delicate Surgical Scissors, 11.5 cm Cat No. 05-0320.

3. Fine round-end forceps, e.g., Graefe forceps—straight and curved.

4. Fine curved forceps, e.g., Student Dumont #7 Forceps—Curved.

5. Self-closing fine forceps.

6. Microcautery, e.g., Bovie, Aaaron Medical, "Change-A-Tip" High Temperature Microcautery.

2.5 Anesthetic

1. Option 1; injectable ketamine hydrochloride (150–180 mg/kg) and xylazine hydrochloride (10 mg/kg) made up together in saline (both at 10 mg/mL). Anesthesia is induced by intraperitoneal injection of the mixture.

2. Option 2; inhaled anesthetic, e.g., sevoflurane (trade name *Sevorane*).

3 Methods

3.1 Mouse Anesthesia and Jugular Vein Surgery

1. Anesthetize mouse. A commonly used anesthetic is a mixture of ketamine and xylazine (*see* Subheading 2.5) delivered intraperitoneally. Allow 5–10 min for the mouse to be completely anesthetized before starting any surgery. A paw pinch test is useful to ascertain the level of anesthesia.

2. Intravenous access. To enable intravenous delivery of reagents such as antibodies, fluorescent markers, and additional anesthetic, secure intravenous access is desirable. This is readily achieved via cannulation of either the jugular vein or tail vein. Here we provide a description of the jugular vein surgery.

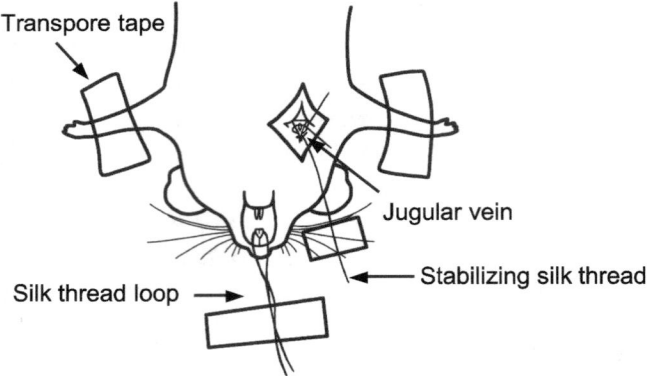

Fig. 2 Setup for cannulation of the mouse jugular vein. Mouse is anesthetized and secured in a supine position to a plexiglas board using transpore tape. The head is immobilized by taping a loop of suture material around the teeth and securing the suture to the board. The jugular vein is exposed via an incision, separated from the surrounding tissue via blunt dissection and cannulated with a saline-filled PE-10 cannula as described. The cannula is held in place with two tightly knotted loops of suture material around the vessel

3. Secure anesthetized mouse on its back, taping down the fore-paws and tail to a plexiglass surgical board. Immobilize the head via a suture looped around the teeth taped down to the surgical board (Fig. 2). To expose the right jugular vein, make a 10 mm incision in the lateral neck skin, directly above the jugular vein starting at the level of the clavicle, extending toward the head. The jugular vein sits under a layer of connective tissue and fat under the skin.

4. Using fine, round-ended forceps (e.g., Graefe forceps), gently separate the fat and connective tissue layers to expose the jugular vein, taking care not to disrupt any blood vessels encountered. Use the forceps to clear a path on either side and underneath the vein and pass two 8–10 cm lengths of 4-0 silk suture under the vein. Tie off the distal end of the vein and use the suture to immobilize the vessel by taping the suture to the operating surface. Tie the remaining proximal suture but do not tighten.

5. Use a 30G needle to make a hole in the uppermost surface of the vein and pass the saline-filled cannula directly under the needle into the lumen of the vein. Pass the tip of the cannula ~5 mm past the second suture and tie a knot in the suture enclosing both the cannula and vein. Ensure the vein is placed correctly by gently drawing back on the syringe. Blood should flow into the cannula. If not, reposition the cannula until this is achieved. Secure the suture with the proximal tie with a double

knot and also double knot the suture at the distal end of the vein over the vessel and the cannula, providing additional support securing the cannula in place (*see* **Note 3**).

3.2 Exteriorization of the Cremaster Muscle

1. Mount the mouse on the viewing platform such that the cremaster muscle can be easily extended over the coverslip in the viewing platform, in the illumination path of the microscope. Hold mouse in position on platform via tape (3M Transpore) or ligatures on the leg. Using small surgical scissors, gently open the skin above the cremaster muscle, making sure to make no contact with the muscle layer beneath the skin. The opening in the skin should extend from the base of the scrotum to adjacent to the penis. Once the skin is open, start the superfusion buffer running over the exposed area (*see* **Note 4**). Each testis is surrounded by an individual cremaster muscle. Select the cremaster immediately below the skin incision and use fine, round-ended Graefe forceps to carefully dissect the muscle away from the surrounding connective tissue and fat (*see* **Note 5**). Connective tissue and fat can be removed from the muscle without the need for incision or cautery. However, it is critical at this point that direct contact and handling of the muscle be minimized.

2. Once the muscle is free of surrounding tissue (it should have a sac-like appearance), use self-closing fine forceps to grasp a small region of muscle tissue at the very base of the muscle, placing the forceps on the platform surface when the muscle is secured. Pass a length of 4-0 suture around the tips of the forceps and make a knot around the region of tissue, taking care to ensure that the knot does not include the tips of the forceps, but is positioned only minimally past the end of the forceps. Tie the suture in place with two knots and release the self-closing forceps. This first suture can now be used to gently move the muscle around in order to remove any remaining connective tissue or fat. Using the suture attached to the base of the muscle, gently stretch the muscle away from the mouse over the underlying coverslip, and tape the suture in place (*see* **Note 6**).

3. Opening the muscle via cautery. Stop the flow of superfusion buffer and use a microcautery to cauterize a 2–3 mm hole through the muscle layer into the inside of the muscle sac. To reduce damage from excessive heating of the tissue, reapply superfusion buffer to the cauterized area immediately after completing the cautery. Pass tips of fine curved forceps through hole and into the testis. Stop the superfusion buffer. Allow the forceps tips to come apart and cauterize vertically along the middle upper surface of the muscle between the forceps tips, reaching the top of the testis. Reapply superfusion buffer to the

cauterized edges of the muscle. The muscle sac should now be open and the testis visible.

4. Using the same technique with self-closing forceps as in **Step 2**, attach a suture to each of the cauterized edges of the muscle. In each case, use the forceps to grasp a small (1–2 mm) piece of the edge of the muscle and secure the suture just beyond the forceps tips. Secure the suture to the muscle with two knots, taking care to place the knots just beyond the end of the forceps. Cut the end of the suture overhanging the muscle immediately adjacent to the knot, leaving the other end of the suture extending away from the muscle. Repeat on the opposite side of the muscle. Secure sutures onto the surface of the platform with tape, and use them to gently stretch the muscle laterally, pulling the two sutures in opposing directions.

5. Removing the testis from the field of view. The testis is connected to the cremaster muscle by a membrane on the bottom of the epididymis. Grasp the testis with forceps and lift it away from the underlying muscle. Use the microcautery to divide the membrane that attaches the testis to the muscle. Take extreme care to avoid touching the muscle with the cautery unit. When the testis is free, use forceps to push it into the abdominal cavity, leaving the muscle free for transillumination. For optimal extension of the muscle, apply one more suture to each side of the muscle (Fig. 3), using the same technique as above. Use all five sutures to gently extend the

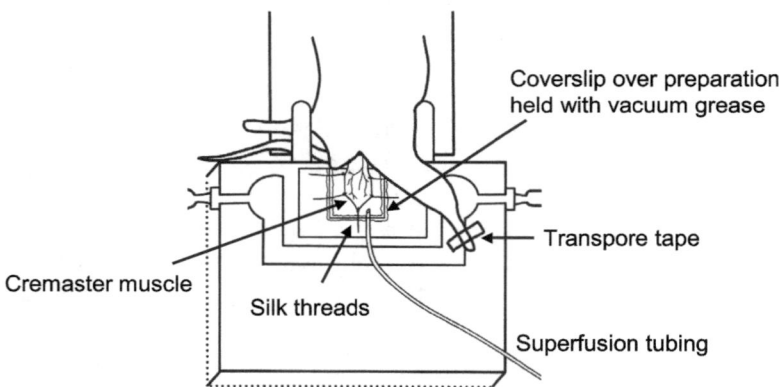

Fig. 3 Cremaster muscle exteriorization. Diagram showing the position of the mouse cremaster muscle on the raised platform following exteriorization. The left leg is secured in place over one leg holder (held in place with tape), while the tail and right leg are secured under the other leg holder. The prepared cremaster muscle is held in position and extended over the visualization window with five pieces of suture material secured to the outer margins of the muscle, as shown. A continuous flow of superfusion fluid is delivered to the muscle via fine tubing connected to a peristaltic pump. The preparation is covered with a coverslip held in place with vacuum grease

muscle over the underlying coverslip, taking care not to over-stretch the muscle and impede blood flow. Cover muscle with superfusion buffer and cover the preparation with a 22 mm × 22 mm coverslip, held in place to the underlying coverslip with vacuum grease around the periphery of the upper coverslip, not contacting the muscle. Ensure buffer is retained underneath the coverslip, without bubbles (*see* **Notes 7** and **8**).

3.3 Imaging the Cremaster Muscle

1. For analysis of leukocyte-endothelial cell interactions, postcapillary venules must be identified. These are readily identified by the direction of blood flow (from small to larger vessels) and, in most mouse strains, by their capacity to support rolling interactions with leukocytes. Interacting leukocytes should be visible as clear circular structures interacting with the venular wall, moving at a significantly slower speed than the other cells carried in the bloodstream. Leukocyte-endothelial cell interactions are assessed in venules of diameters in the range of 20–35 μm.

3.4 Induction of Inflammation

The cremaster muscle is amenable to assessment of both acute and longer-term inflammatory responses.

1. Rapid leukocyte responses can be induced by superfusion of the cremaster preparation with chemoattractants such as CXCL1 or fMLP, with increases in leukocyte adhesion occurring within 15–30 min, and robust transmigration responses apparent within 45–60 min [16, 17].

2. Long-term responses can be induced by injecting proinflammatory cytokines such as IL-1β or TNF subcutaneously in the scrotum adjacent to the cremaster muscle, leading to induction of local leukocyte recruitment that can be assessed 3–5 h later [7] (*see* **Note 9**).

3.5 Analysis of Leukocyte-Endothelial Cell Interactions

In order to study leukocyte-endothelial interactions in postcapillary venules of the cremaster muscle, one to three venules are typically studied, with the same region of venule observed over the time course of the experiment. Recordings are taken at various time points and leukocyte interaction parameters determined off-line during playback analysis. Typically, interactions are observed in a 100-μm length of vessel to standardize results. The following parameters are assessed:

1. Leukocyte rolling flux: The number of leukocytes that are rolling/interacting within the vessel, e.g., traveling more slowly than the erythrocytes in the blood. This is determined by counting the number of leukocytes that roll past a defined point in a minute (cells/min).

2. Leukocyte rolling velocity: Rolling velocity is calculated by (1) measuring the time it takes for a leukocyte to roll the 100-μm defined distance, and calculating the velocity as d/t, where d = distance = 100 μm and t = time taken to roll this distance. Typically, velocity is determined for ~20 cells per vessel and data expressed as (μm/s).

3. Leukocyte adhesion: Leukocytes are counted as adherent or static if they remain stationary for 30 s or longer and are assessed within the defined 100 μm length of vessel.

4. Leukocyte crawling: Leukocyte intravascular crawling is a step in the leukocyte recruitment cascade that occurs after firm adhesion but before transmigration. Cells flatten, extend pseudopods and crawl to sites of transmigration. Crawling cells can be tracked using time-lapse video microscopy. Distance (μm), speed (μm/min), and the direction of cell crawling within the vessels can be calculated from when they first adhere until they transmigrate or detach. Image J (FIJI) tracking analysis can be used to quantitate these parameters.

5. Leukocyte emigration (also known as transmigration). Leukocyte emigration is defined as the number of leukocytes visible outside of the post capillary venule per microscopic field of view.

4   Notes

1. Typically, freshly prepared bicarbonate buffer requires adjustment of the pH to 7.4, using high concentration HCl. This is best done immediately before use in experiments.

2. The mouse cremaster preparation is amenable for imaging via a wide range of modalities. The bulk of the imaging experiments examining the role of MIF were performed using wide-field transillumination microscopy, with images recorded using an analog VHS video system in either color or monochrome. These recordings were performed at video rate (24 frames per second—fps) typically recording fields featuring a single post-capillary venule for approximately 3 min. This approach capitalizes on the fact that no exogenous labels or reporter molecules are necessary to visualize leukocytes in transilluminated preparations. The inherent difference in contrast between rolling and adherent leukocytes relative to noninteracting immune cells and other rapidly moving particles in the bloodstream means that these cells are readily visible via transillumination. However, the subsequent advent of digital, confocal, and multiphoton imaging means that there are now a

wide range of options for recording data from cremaster muscle preparations:

Digital video: Recent developments in camera technology, such as CMOS (complementary metal–oxide semiconductor) cameras, now provide a combination of high speed (e.g., 30 fps), high resolution (e.g., 1024×1024), and excellent sensitivity that better anything previously achievable using an analog system. CMOS cameras can be mounted on a conventional transillumination microscope as for analog cameras, but require operation via appropriate image acquisition software run via a high-quality computer.

Confocal microscopy: Leukocyte endothelial cell interactions can also be assessed with confocal microscopy [18, 19]. In contrast to transillumination microscopy, confocal microscopy is a laser-based form of microscopy that works via epifluorescence. As such it requires incorporation of fluorescently labeled markers for detection of target cells. For analysis of leukocyte-endothelial cell interactions, leukocytes are labeled either by exogenous stains (e.g., PE-anti-Ly6G to stain neutrophils) or use of specific immune cell reporter mice (e.g., LysM-eGFP to detect neutrophils). These are typically combined with a fluorescent vascular label to enable visualization of the vasculature and the immune cell of choice.

Both scanning confocal microscopy [18] and spinning disk confocal microscopy [19] have been used for analysis of leukocyte trafficking in the cremaster muscle, with the latter format typically being capable of achieving higher rates of image capture. However, developments such as resonant scanners have markedly increased the frame rate able to be achieved by scanning confocal systems such that this is no longer a limitation.

Multiphoton imaging: Multiphoton microscopy is recognized for its capacity to visualize structures below the surface of tissues. This modality has made an enormous impact in analysis of organs such as the lymph node, brain, and kidney [20–22]. However, it can also be applied to intravital microscopy of the cremaster muscle. As for confocal microscopy, multiphoton imaging is laser-based and requires use of fluorochromes and/or fluorescent reporter proteins to label cells and structures of interest, with a few exceptions. It is particularly amenable for three-dimensional analysis of the cremasteric microvasculature.

3. As an alternative, a tail vein cannula can be generated using the tip of a 30G needle secured in the end the PE10 cannula. Insert this in the vein, using a temporary ligature to inflate the tail vein. Once backflow of blood into the cannula is achieved, tape the cannula securely to the tail.

4. As soon as muscle is exposed, it is important that steps are taken to ensure that the tissue does not dry out. We achieve this by delivering a continuous flow of superfusion buffer (heated to 37 °C) over the muscle, via fine silicone tubing, during all steps of the surgery, apart from those involving cauterization. When using cautery, it is necessary to stop the flow of superfusion buffer to enable the cautery to work efficiently. However, as soon as practical after completing cautery steps, buffer superfusion should be recommenced.

5. When dissecting the cremaster muscle, it is important to differentiate the muscle from connective tissue and fat. Connective tissue tends to have a shiny appearance and can be gently stretched and torn without resulting in bleeding. Fat has an opaque, white appearance and typically can be removed from the area without requirement for cutting or cautery. The muscle of interest is of a tan slightly opaque color. In order to achieve a high-quality preparation, it is important to limit the amount of handling the muscle is subjected to.

6. At this point it is important to ensure the testis is not rotated from its natural position because the major blood vessels supplying the muscle are located at the rear. If the muscle is rotated, these major vessels could potentially be disrupted during the cautery step. To avoid this, before taping down the first suture, allow the muscle to relax to its original position and then stretch it out again. This should ensure the vessels remain at the rear of the muscle, and less at risk of damage via the microcautery.

7. For clear imaging, we find it useful to place a coverslip on the top of the preparation. This facilitates retention of a consistent amount of superfusion of buffer on the muscle, and ensures optical clarity for microscopy. This approach is suitable for use with both dry and immersion lenses, in that immersion fluids will be easily retained on the uppermost coverslip.

8. Typically, it is ideal to delay imaging for 20–30 min after completing the surgery, to allow endothelial activation stemming from surgical preparation of the tissue to abate.

9. Injection of inflammatory mediators (LPS or TNF) directly into the scrotum of mice was used to compare responses in wild-type and MIF-deficient animals, revealing a role for endogenous MIF in promoting leukocyte recruitment [7]. In addition, local application of exogenous MIF has been shown to induce leukocyte recruitment over a similar time scale [8, 9]. These experiments revealed a capacity of MIF to induce local expression of CCL2 and thereby facilitate monocyte/macrophage recruitment.

Acknowledgements

This work is supported by funding from the National Health and Medical Research Council of Australia (1042775, MJH) and the Monash University School of Clinical Sciences at Monash Health (MUN).

References

1. Calandra T, Roger T (2003) Macrophage migration inhibitory factor: a regulator of innate immunity. Nat Rev Immunol 3 (10):791–800
2. Ayoub S, Hickey MJ, Morand EF (2008) Mechanisms of disease: macrophage migration inhibitory factor in SLE, RA and atherosclerosis. Nat Clin Pract Rheumatol 4(2):98–105
3. Bernhagen J et al (1996) An essential role for macrophage migration inhibitory factor in the tuberculin delayed-type hypersensitivity reaction. J Exp Med 183:277–282
4. Lan HY et al (1997) The pathogenic role of macrophage migration inhibitory factor in immunologically induced kidney disease in the rat. J Exp Med 185(8):1455–1465
5. Leech M et al (1998) Involvement of macrophage migration inhibitory factor in the evolution of rat adjuvant arthritis. Arthritis Rheum 41(5):910–917
6. Hoi AY et al (2006) Macrophage migration inhibitory factor deficiency attenuates macrophage recruitment, glomerulonephritis, and lethality in MRL/lpr mice. J Immunol 177 (8):5687–5696
7. Gregory JL et al (2004) Reduced leukocyte-endothelial cell interactions in the inflamed microcirculation of macrophage migration inhibitory factor-deficient mice. Arthritis Rheum 50:3023–3034
8. Gregory JL et al (2006) Macrophage migration inhibitory factor induces macrophage recruitment via CC chemokine ligand 2. J Immunol 177(11):8072–8079
9. Bernhagen J et al (2007) MIF is a noncognate ligand of CXC chemokine receptors in inflammatory and atherogenic cell recruitment. Nat Med 13(5):587–596
10. Stark K et al (2013) Capillary and arteriolar pericytes attract innate leukocytes exiting through venules and 'instruct' them with pattern-recognition and motility programs. Nat Immunol 14(1):41–51
11. Ley K et al (2007) Getting to the site of inflammation: the leukocyte adhesion cascade updated. Nat Rev Immunol 7(9):678–689
12. Hickey MJ, Westhorpe CL (2013) Imaging inflammatory leukocyte recruitment in kidney, lung and liver – challenges to the multi-step paradigm. Immunol Cell Biol 91(4):281–289
13. Cheng Q et al (2010) Macrophage migration inhibitory factor increases leukocyte-endothelial interactions in human endothelial cells via promotion of expression of adhesion molecules. J Immunol 185(2):1238–1247
14. Santos LL et al (2011) Macrophage migration inhibitory factor regulates neutrophil chemotactic responses in inflammatory arthritis in mice. Arthritis Rheum 63(4):960–970
15. Fan H et al (2011) Macrophage migration inhibitory factor and CD74 regulate macrophage chemotactic responses via MAPK and Rho GTPase. J Immunol 186(8):4915–4924
16. Wee JL et al (2015) Tetraspanin CD37 regulates b2 integrin-mediated adhesion and migration in neutrophils. J Immunol 195 (12):5770–5779
17. Nicholls AJ et al (2018) Activation of the sympathetic nervous system modulates neutrophil function. J Leukoc Biol 103(2):295–309
18. Woodfin A et al (2011) The junctional adhesion molecule JAM-C regulates polarized transendothelial migration of neutrophils in vivo. Nat Immunol 12(8):761–769
19. Vanheule V et al (2017) CXCL9-derived peptides differentially inhibit neutrophil migration in vivo through interference with glycosaminoglycan interactions. Front Immunol 8:530
20. Mempel TR, Henrickson SE, von Andrian UH (2004) T-cell priming by dendritic cells in lymph nodes occurs in three distinct phases. Nature 427:154–159
21. Devi S et al (2013) Multiphoton imaging reveals a novel leukocyte recruitment paradigm in the glomerulus. Nat Med 19:107–112
22. Snelgrove SL et al (2017) Activated renal dendritic cells cross present intrarenal antigens after ischemia reperfusion injury. Transplantation 101:1013–1024

Chapter 4

Analysis of Secreted MIF from Cultured Murine Bone Marrow-Derived Immune Cells

Tali Lang

Abstract

Macrophage migration inhibitory factor (MIF) is expressed and released ubiquitously by numerous cell types and tissues. MIF is detected and constitutively expressed at the protein level both intra- and extracellularly. This chapter outlines methods for cultivating, purifying, detecting, and quantifying concentrations of MIF from murine primary derived macrophages and dendritic cell culture supernatants.

Key words MIF, BMM, BMDC, Cell culture, ELISA, Western blot analysis

1 Introduction

Macrophage migration inhibitory factor (MIF) is a pleiotropic cytokine that regulates both innate and adaptive immune responses [1]. MIF was originally discovered as a T cell derived cytokine [2, 3] but is now known to be expressed by multiple immune cell types including monocytes, macrophages, dendritic cells, lymphocytes, eosinophils, and neutrophils [4]. MIF is a 12.5 kDa protein comprising 114 amino acids, and is highly conserved across species, with murine and human MIF showing 90% homology [5]. MIF is constitutively expressed and released into the extra-cellular space following stimulation by both pathogen-associated molecular patterns (PAMPs) and danger associated molecular patterns (DAMPs). Herein we describe the cultivation of murine derived macrophages and dendritic cells, as well as the purification and preparation of cellular supernatants required for the accurate detection and analysis of both intracellular and secreted MIF using commercial assays such as ELISA and Western blot analysis.

James Harris and Eric F. Morand (eds.), *Macrophage Migration Inhibitory Factor: Methods and Protocols*,
Methods in Molecular Biology, vol. 2080, https://doi.org/10.1007/978-1-4939-9936-1_4,
© Springer Science+Business Media, LLC, part of Springer Nature 2020

2 Materials

2.1 Mice

All animals should be bred in the same specific pathogen free (SPF) facility on the same genetic background, age and gender matched.

2.2 Making L-Cell (M-CSF) Conditioned Media

1. L929 cells.
2. Complete RPMI medium: Roswell Park Memorial Institute 1640 (RPMI 1640) medium supplemented with 10% fetal calf serum (FCS), Penicillin and Streptomycin (100 U/mL and 100 µg/mL, respectively) and L-glutamine (100 µg/mL).
3. T175 tissue culture flasks.
4. 50 mL syringes.
5. 0.45 µm syringe filters.
6. 50 mL tubes.

2.3 Making GM-CSF Conditioned Media

1. Ag8653 cells.
2. Complete RPMI medium (*see* Subheading 2.2).
3. T175 tissue culture flasks.
4. 50 mL syringes.
5. 0.45 µm syringe filters.
6. 10 mL and 50 mL plastic tubes.

2.4 Culturing of Primary Derived Bone Marrow Macrophages (BMM)/ Dendritic Cells (BMDC)

1. Wild-type mice (C57BL/6), 8–12-week-old.
2. 70% Ethanol (v/v) in distilled H_2O (dH_2O).
3. Red blood cell lysis buffer; 155 mM NH_4Cl, 12 mM $NaHCO_3$, 0.1 mM EDTA in Milli-Q H_2O. Filter sterilize with 0.2 µm syringe-tip filter. Commercially available red blood cell lysis solutions can also be used.
4. Complete RPMI medium (*see* Subheading 2.2).
5. M-CSF or GM-CSF conditioned medium (*see* Subheadings 2.2 and 2.3).
6. Cell dissociation reagent (e.g., Trypsin/EDTA).
7. 0.4% (w:v) Trypan blue stock solution.
8. 3 mL syringes.
9. 15 mL polypropylene tube, sterile.
10. 23G and 26G needles, sterile.
11. 6-well cell culture plates.
12. T175 bacteriological culture flasks for BMM, T175 tissue culture flask for BMDC (*see* **Note 1**).
13. Dissection tools: scissors, forceps.
14. Hemocytometer/cell counter.

2.5 Culturing of Monocyte/ Macrophage Cell Lines

Examples of commonly used monocyte/macrophage cell lines include murine RAW267.4 cells, immortalized bone marrow-derived macrophages (iBMM) and J774 cells, and human THP-1 cells. In most cases the same medium can be used for each (*see* **Note 2**).

1. Sterile Phosphate Buffered Saline: 137 mM Nacl, 2.7 mM KCl, 10 mM Na_2HPO_4, 1.8 mM KH_2PO_4 in Milli-Q H_2O, pH 7.4. Autoclave or filter to sterilize.

2. Cell dissociation reagent (e.g., Trypsin/EDTA).

3. Complete DMEM: Dulbecco's Modified Eagle's Medium supplemented with 5% heat-inactivated FCS, 2 mM L-glutamine, 100 U/mL penicillin, 100 μg/mL streptomycin (*see* **Note 3**).

4. 0.4% (w:v) Trypan blue solution.

5. Sterile 50 mL polypropylene tubes.

6. 96-well cell culture flat bottom plates, 6-well cell culture plate, 10 cm round plastic plates.

7. Hemocytometer/cell counter.

2.6 Purification of MIF Protein from Cultured Cell Supernatants

1. Methanol.

2. Chloroform.

3. 5× Laemmli buffer: 0.3125 M Tris–HCl, pH 6.8, 50% (v:v) Glycerol, 10% (w:v) SDS 0.02% (w:v) Bromophenol blue, 10% (v:v) reducing agent (2-mercaptoethanol or DTT (dithiothreitol), added immediately before use) in Milli-Q H_2O.

4. 1.7 mL collection tubes.

2.7 Cytokine Quantification Using Commercial ELISA Kits

1. Commercial or in-house ELISA kits or antibody pairs. Where not included in the kit, the following may also be needed: TMB (3,3′,5,5′-tetramethylbenzidine) substrate solution (*see* **Note 4**), stop solution; 0.2–2M H_2SO_4 (*see* **Note 5**).

2. Wash solution: 0.05% Tween® 20 in PBS, pH 7.4.

3. High-binding flat-bottom 96-well ELISA/RIA plates.

4. Microplate reader set at appropriate wavelength.

5. 10–300 μL multichannel pipette.

2.8 Detection of MIF by Western Blot Analysis

1. 12–15% SDS-PAGE gels.

2. Polyvinylidene difluoride (PVDF) membrane.

3. Running Buffer (*see* **Note 6**).

4. TBS-T: 50 mM Tris–Cl, 150 mM NaCl, pH 7.5 in Milli-Q H_2O.

5. Block buffer: 5% skimmed milk (w:v) in TBS-T.

6. Chemiluminescence (ECL) Western blotting reagents.

7. X-ray film or luminescence reader.

3   Methods

3.1   Making M-CSF Conditioned Medium

1. Seed L929 ("L-cells") at a final concentration of 4.7×10^5 in a T75 cm cell culture flask containing 55 mL complete RPMI medium.

2. Culture cells for 7 days in a humidified incubator at 37 °C, 5% CO_2.

3. Collect supernatant and centrifuge at $300 \times g$ for 5 min to ensure any cells are pelleted at bottom of tube.

4. Transfer supernatant and filter through a 0.45 µm filter

5. Store in 50 mL aliquots at −20 °C.

3.2   Making GM-CSF Conditioned Medium

1. Seed 4×10^7 cells in 50 mL of complete RPMI medium (RPMI 1640 supplemented with 10% heat-inactivated fetal calf serum (FCS), 2 mM glutamine, 100 U/mL penicillin, 100 µg/mL streptomycin) in a T75 cell culture flask.

2. Culture cells for 4 days in a humidified incubator at 37 °C, 5% CO_2.

3. Collect supernatant and centrifuge at $300 \times g$ for 5 min to ensure any cells are pelleted at bottom of tube.

4. Transfer supernatant and filter through a 0.45 µm filter. Store in 10 mL aliquots at −20 °C.

3.3   Isolation and Cultivation of BMM/BMDC

1. Euthanize mice either by CO_2 inhalation or cervical dislocation immediately prior to the procedure.

2. In a fume hood, disinfect the skin with 80% ethanol.

3. Use forceps to hold up the skin on the mouse's thigh area and cut open the skin, exposing the muscles of the leg.

4. Cut away major muscles surrounding femur so as to expose bone toward the knee joint and hip joint is exposed.

5. Use scissors to cut through the hip to release the whole leg (femur and tibia) while keeping the femur intact.

6. Cut tibia below knee joint; remove tissue around tibia and cut feet off at a point low enough not to compromise tibia.

7. Place bones in 5 mL complete RPMI in a 6-well plate on ice.

8. Take bones from fume hood into a cell culture laminar flow hood for the remainder of the processing (*see* **Note** 7).

9. Set up 1 × 15 mL tube for each mouse with fresh sterile complete RPMI (5 mL total volume)

10. Load a 3 mL syringe with sterile complete RPMI medium and attach a 26-gauge needle.

11. Hold the femur bone with forceps and cut off the ends with a pair of fine scissors.

12. Flush both femur and tibia with complete RPMI (the whole tibia/femur should be intact) by inserting the needle into the bone marrow cavity and flush out the bone marrow cells (into the prepared 15 mL tube).

13. Centrifuge cells at $300 \times g$ for 5 min at room temperature and discard supernatant.

14. Add 1 mL of red cell lysis buffer, resuspend pellet, and incubate at room temperature for 7 min.

15. Add 9 mL of complete RPMI medium and mix.

16. Centrifuge cells at $300 \times g$ for 5 min. Discard supernatant and remove any fat/muscle (*see* **Note 8**).

17. Resuspend cell pellet in 10 mL complete RPMI.

18. Mix 10 μL of cell suspension with 10 μL Trypan blue. Load 10 μL into a chamber of the hemocytometer. Count viable cells.

19. To make BMM, plate cells (25 mL) at 1×10^6 cells/mL in T175 bacteriological culture flasks in complete RPMI medium supplemented with 10–20% M-CSF conditioned medium (*see* **Note 9**).

20. On day 3, add 15 mL complete RPMI medium supplemented with M-CSF conditioned medium.

21. Harvest cells on days 6–8, using cell dissociation buffer to remove the adherent BMM. Centrifuge cells at $300 \times g$ for 5 min and resuspend cells in 10 mL complete medium supplemented with M-CSF conditioned medium. Count cells and plate for use.

22. To make BMDC, plate cells in T175 tissue culture flasks supplemented with 10–20% GM-CSF conditioned medium (*see* **Note 9**).

23. On Day 3, add 15 mL complete RPMI medium supplemented with GM-CSF conditioned medium.

24. On day 6, remove non-adherent cells by removing the medium and gently washing once with complete RPMI medium or PBS. Culture cells in 25 mL complete RPMI medium supplemented with GM-CSF conditioned medium.

25. On day 9, add 15 mL complete RPMI medium supplemented with GM-CSF conditioned medium.

26. On day 11, collect the non-adherent and loosely adherent BMM by gently aspirating the cells with their culture medium. Centrifuge the cells at $300 \times g$ for 5 min, resuspend in 10 mL

complete RPMI medium supplemented with GM-CSF conditioned medium. Count cells and plate for use.

3.4 Cell Culture for Quantification of Secreted MIF in Cellular Supernatants for Detection by ELISA

1. Seed cells of choice at $0.5-1 \times 10^6$ cells/mL (200 μl/well) in 96-well tissue culture plates overnight.

2. The following day treat/stimulate cells as required. $300 \times g$ for 5 min at room temperature.

3. Using a multichannel pipette, remove cell culture supernatants and transfer to another 96-well plate for immediate use or storage at $-80\ °C$ for future analysis.

3.5 Quantification of Secreted MIF in Cell Culture Supernatants Using ELISA

1. Before commencing experiments, it is important to first optimize conditions for running your samples in each commercial ELISA kit that you use. Try a range of sample dilutions to ensure the concentrations of MIF lie within the detection limits of the ELISA (*see* **Note 10**). Avoid freeze/thaw cycles of your samples as this effects MIF protein stability and detectability.

3.6 Cultivation and Purification of Cell Culture Supernatants for Detection by Western Blot Analysis

1. Seed cells of choice at 2×10^6 cells/well in 6-well culture plates (2 mL total volume) overnight.

2. The next day stimulate/treat cells as required in RPMI 1640 medium containing 1% FCS (up to 6 h treatment, if longer treatment periods are required use at least 5% FCS).

3. Collect cell supernatants and divide them for processing into 500 μL aliquots in 1.7 mL collection tubes (*see* **Note 11**).

4. To every 500 μL aliquot of cell supernatant, add 500 μL of methanol and 100 μL chloroform, vortex well. You should see cloudiness appearance.

5. Centrifuge tubes in a microfuge at 13,000 rpm for 3 min.

6. The protein is at the interface as a solid white layer; remove and discard the upper phase (methanol/H_2O).

7. Wash the remaining protein/chloroform with 500 μL methanol. Vortex, centrifuge tubes in a microfuge at 13,000 rpm for 3 min.

8. The protein should be clearly pelleted at the bottom of the tube. Completely remove the supernatant.

9. Air-dry the pellet in the fume hood for 15 min.

10. Resuspend pellet in at least 30 μL of $1\times$ Laemmli buffer + reducing agent (*see* **Notes 12** and **13**). At this point, samples can be stored $<-20\ °C$ until analysis.

11. Boil samples for 5–10 min at $95\ °C$.

3.7 Detection of MIF in Cell Culture Supernatants by Western Blot Analysis

1. Run 20 µL of purified protein samples collected from cell culture supernatants or cell lysates on a 15% SDS-PAGE gel for 30 min at 90 V, then increase voltage to 120 V until adequate separation is acheived.

2. Transfer onto PVDF membrane.

3. Block membranes in order to prevent nonspecific binding with block buffer for 1 h at room temperature.

4. Incubate membranes with primary antibodies diluted in block buffer on a roller overnight at 4 °C (*see* **Note 14**).

5. Wash 3 × 15 min on a roller at room temperature in TBST

6. Incubate membrane with horseradish peroxidase-conjugated secondary antibody diluted in block buffer for 1 h at room temperature.

7. Wash 3 × 30 min in TBST.

8. We find MIF is best detected using enhanced chemiluminescence (ECL) Western blotting reagents.

9. Develop images on X-ray film or using a luminescent reader.

4 Notes

1. It is important not to use tissue culture plastics for culturing of BMM, as they will adhere very strongly, such that removing them requires high concentrations of trypsin which can be detrimental to the cells. For BMDC cell culture, cell culture plastic is required, as non-adherence to these is a defining feature of the required cells.

2. In our laboratory we routinely cultivate immortalized cell lines in complete DMEM growth medium; however, other groups prefer and have published cultivation of these cell lines in other growth media such as RPMI-1640. It is recommended that you optimize cell growth culture conditions for your cell line of interest.

3. Other media may be used if preferred (e.g. complete RPMI medium).

4. Other substrates for HRP can alternatively be used, including ABTS (2,2′-Azinobis [3-ethylbenzothiazoline-6-sulfonic acid]-diammonium salt), OPD (*o*-phenylenediamine dihydrochloride).

5. Check the manufacturer's instructions supplied with the ELISA kit for concentration of stop solution.

6. The running buffer required will depend on the type of gel used (e.g., Tris-Glycine, Bis-Tris, Tris-acetate). Check manufacturer's recommendation.

7. Injury with sharps (scissors, syringes, needles) can lead to possible exposure to infectious biological materials, so ensure suitable personal protective equipment (PPE) such as latex gloves, lab coat, face masks, and protective eye glasses are worn.

8. Cell suspensions can be passed through cell strainers to remove large pieces of fat/tissue.

9. The concentration of M-CSF and GM-CSF conditioned medium required for optimum cell growth and differentiation can vary and we suggest optimizing final concentrations (commonly 10–20%) before for each batch. This can be aided by testing M-CSF/GM-CSF concentrations in conditioned media using an ELISA. As an alternative to conditioned media, recombinant M-CSF or GM-CSF can be added at final concentrations of 10–25 ng/mL.

10. It is recommended that you run samples neat, as well as perform a 1:2 serial dilution of both control and treated samples to ascertain best dilution for use in your experiments.

11. At this stage you should also process cellular lysates as a control to see the levels of intracellular MIF in both untreated and treated samples by Western blot analysis. Two approaches which can be taken are (1) detection of MIF in a crude whole cell lysate or (2) specific pull-down approach for detection of MIF and binding partners through conjugation of a MIF antibody onto Protein G agarose beads.

12. If you are combining resuspended protein pellets from multiple tubes, first resuspend each pellet in dH_2O (40 μL), then add $5\times$ Laemmli buffer (10 μL) to have a total volume no greater than 50 μL.

13. 2ME is highly toxic and inhalation, contact with the skin or eyes should be avoided. Preparation and addition of 2ME (50 μL) to $5\times$ Laemmli buffer (950 μL) should be carried out in a chemical hood, while wearing gloves, safety glasses, and a protective lab coat.

14. There are many commercial antibodies available for detection of MIF. We recommend optimization based on your cells/species of choice. Where possible, we highly recommend MIF knockout-validated antibodies.

References

1. Lang T et al (2015) MIF: implications in the pathoetiology of systemic lupus erythematosus. Front Immunol 6:577

2. Bloom BR, Bennett B (1966) Mechanism of a reaction in vitro associated with delayed-type hypersensitivity. Science 153(3731):80–82

3. David JR (1966) Delayed hypersensitivity in vitro: its mediation by cell-free substances formed by lymphoid cell-antigen interaction. Proc Natl Acad Sci U S A 56(1):72–77

4. Baugh JA, Bucala R (2002) Macrophage migration inhibitory factor. Crit Care Med 30 (1 Supp):S27–S35

5. Calandra T, Roger T (2003) Macrophage migration inhibitory factor: a regulator of innate immunity. Nat Rev Immunol 3(10):791–800

Chapter 5

Measuring MIF in Biological Fluids

Fabien B. Vincent and Tali Lang

Abstract

MIF is a key regulator of host immune responses and increased levels secreted from cells, or found circulating systemically, have been implicated in the pathogenesis of many inflammatory and autoimmune disorders. Here, we describe methods for detecting and quantifying extracellular concentrations of MIF in both human- and murine-derived biological samples.

Key words MIF, Serum, Plasma, Urine, Cell culture, ELISA, Multiplex immunoassay, Quantibody®

---

1 Introduction

Macrophage migration inhibitory factor (MIF) is a pleiotropic cytokine that regulates host immune responses through a broad range of immunomodulatory properties [1]. MIF was originally discovered as a T cell-derived cytokine [2, 3]; however, extensive research over the past 20 years has shown that MIF is released from multiple cell types and tissues [4]. Moreover, MIF has been shown to act not only as a cytokine but also as a growth factor and a stress factor released by cells of the anterior pituitary gland [5].

MIF is a 12.5 kDa protein comprising 114 amino acids, and is highly conserved across species, with murine and human MIF showing 90% homology [6]. MIF is constitutively expressed and stored within intra-cytosolic pools, and is released into the extracellular space following stimulation by Toll-like Receptor (TLR) ligands, mitogens, and other pro-inflammatory cytokines [6]. Moreover, human studies have shown MIF to be present at high concentrations within the bloodstream (healthy adults between 2 and 10 ng/mL), and readily detectable in other biological samples, including urine [6, 7]. These levels are further increased in response to infection, or as a consequence of dysregulated immune responses associated with autoimmune and auto-inflammatory disorders [7–9].

James Harris and Eric F. Morand (eds.), *Macrophage Migration Inhibitory Factor: Methods and Protocols*,
Methods in Molecular Biology, vol. 2080, https://doi.org/10.1007/978-1-4939-9936-1_5,
© Springer Science+Business Media, LLC, part of Springer Nature 2020

Herein, we describe the preparation of cellular supernatants and human biological samples, required for the accurate detection and analysis of secreted MIF using commercial assays such as ELISA and multiplex immunoassays.

2 Materials

2.1 Mice

All animals should be bred/housed in the same specific pathogen free (SPF) facility on the same genetic background, age- and gender-matched.

2.2 TLR-Dependent In Vivo Inflammatory Models

1. Mice, 8–12-week-old male/female.
2. Lyophilized TLR agonists dissolved in sterile endotoxin free water to the following stock concentrations; LPS (1 mg/mL), IMQ (1 mg/mL), or CpG (500 μM). Further dilute solubilized TLR ligands to a working concentration in sterile sodium chloride saline in a 5 mL polypropylene plastic tube; LPS, 50 μg/200 μL; CpG, 50 μg/200 μL; or IMQ, 100 μg/200 μL.
3. Sterile sodium chloride saline (0.9%) for injection (commercially available ampules).
4. 1 mL insulin syringe.
5. 26 G × 12 mm needles.
6. 5 mL polypropylene plastic tube (low adherence).
7. Digital scales.
8. Laminar flow cabinet.

2.3 Collection of Murine Cardiac Puncture Blood

1. 1 mL syringe.
2. 23-25G needles.
3. Sterile 1.7 mL microtubes.
4. Pipettes, 0.5–1 mL, single channel.
5. Refrigerated centrifuge.
6. CO_2 gas (for euthanasia).

2.4 Collection and Processing of Human Serum and Plasma Samples

1. Venepuncture needles.
2. Blood collection tubes (*see* **Note 1**).
3. Refrigerated benchtop centrifuge.
4. Pipettes, 0.010–1 mL, single channel with filter tips.
5. Sterile 0.5 mL and 1.7 mL microtubes.
6. Laminar flow cabinet.

2.5 Collection and Processing of Human Urine Samples

1. Urine collection in yellow lid specimen jar.
2. Refrigerated benchtop centrifuge.
3. Pipettes, 0.010–1 mL, single channel with filter tips.
4. Sterile 0.5 mL and 1.7 mL microtubes.
5. Laminar flow cabinet.

2.6 Cytokine Quantification Using Commercial ELISA Kits

1. Commercial or in-house ELISA kits or antibody pairs. Where not included in the kit, the following may also be needed:
2. Phosphate buffered saline (PBS): 137 mM NaCl, 10 mM Na_2HPO_4, 1.8 mM KH_2PO_4, 2.7 mM KCl in Milli-Q water, adjust to pH 7.4 with HCl. Sterilize by autoclaving or filtration.
3. Wash solution: 0.05% (v:v) Tween® 20 in PBS.
4. TMB (3,3′,5,5′-tetramethylbenzidine) substrate solution (*see* **Note 2**)
5. Stop Solution: . 0.2–2 M H_2SO_4 (*see* **Note 3**)
6. Nunc MaxiSorp™ flat-bottom 96-well plates.
7. Microplate reader set at 450 nm with wavelength correction set at 540 or 570 nm.
8. 0.01–0.5 mL multichannel pipette.

2.7 Quantibody® multiplex ELISA

1. Quantibody® assay (Raybiotech, Norcross, GA, USA).
2. Human serum/plasma samples.
3. Pipettes, 0.010–1 mL, single and multichannel with filter tips.
4. Vortex.
5. Refrigerated benchtop centrifuge.
6. Fluorescence laser scanner.
7. RayBiotech Q-Analyzer software.

2.8 ProcartaPlex® Multiplex Immunoassay

1. ProcartaPlex® immunoassay (Thermo Fisher Scientific).
2. Human serum/plasma samples.
3. Pipettes, 0.010–1 mL, single and multichannel with filter tips.
4. Multichannel pipette reservoir.
5. Vortex.
6. Refrigerated benchtop centrifuge.
7. Plate mixer (500 rpm).
8. Magnetic plate holder.
9. Luminex technology platform (MAGPIX®, Luminex™ 100/200, or FLEXMAP 3D®).

3   Methods

3.1   TLR-Induced In Vivo Challenge

1. Dissolve lyophilized TLR ligands to listed working concentrations in endotoxin -free water to prepare stock solutions in sterile polypropylene plastic tubes (*see* **Note 4**).

2. Pre-fill 1 mL insulin syringes with LPS, IMQ, CpG, or vehicle control saline (200 μL), and attach a 26G needle.

3. In a laminar flow hood, weigh mice, then restrain by scruffing high around neck and shoulders to restrict excessive movement and ensure body is straight and elongated in the palm of your hand.

4. Inject in a total volume of 200 μL into mice intra-peritoneally (at least n = 5/treatment group), of vehicle control (saline), LPS (50 μg/200 μL), CpG (50 μg/200 μL), or IMQ (100 μg/200 μL).

5. Monitor mice every 30 min to ensure mice are not displaying signs of severe sepsis, as per local ethics requirements (*see* **Note 5**).

6. Two hours post-injection, cull mice by CO_2 asphyxiation.

3.2   Collection and Processing of Cardiac Blood

1. Label sterile 1.7 mL microtubes for blood collection.

2. Assemble 1 mL syringes with 23–25G needle.

3. Once the mouse is euthanized, spray fur with 70% ethanol.

4. Collect cardiac blood using dorsal or lateral approach by placing the needle bevel up into the chest and puncture the heart. If the needle is in the heart, blood will start to flow into the syringe.

5. Empty syringe contents into sterile collection microtubes and allow to clot at room temperature for at least 30 min or place samples at 4 °C and store overnight.

6. Remove clot by centrifugation at 1,000–2,000 × g for 15 min at 4 °C.

7. Collect the resulting supernatant (top clear layer), which is the serum component and aliquot into appropriate volumes for storage at −80 °C and subsequent analysis (*see* **Note 6**).

3.3   Collection and Processing of Human Serum

1. Collect 8.5 mL of whole blood by venepuncture into BD Vacutainer® Plastic SST™ II Advance tubes.

2. Sit blood samples at room temperature for at least 30 min to clot.

3. Centrifuge blood tubes at 1,300 × g for 15 min at 4 °C.

4. The top clear layer is the serum component and can be stored in its respective blood collection tube (defined as unseparated) at 4 °C for up to 3 days (*see* **Note 7**).

5. Following centrifugation, aliquot serum (minimum 200 μL) into sterile microtubes and store at −80 °C until required for analysis (*see* **Note 8**).

3.4 Collection and Processing of Human Plasma

1. Collect 10 mL of whole blood by venepuncture into BD Vacutainer® Plastic K2EDTA tubes.

2. Invert tube, 8–10 times back and forth.

3. Sit tubes at 4 °C for at least 30 min.

4. Centrifuge collection tubes at 1,300 × *g* for 15 min at 4 °C.

5. Collect plasma layer and aliquot plasma into sterile microtubes and store at −80 °C until required for analysis (*see* **Note 8**).

3.5 Collection and Processing of Human Urine

1. Collect first morning mid-stream urine sample in a yellow lid urine pot (*see* **Note 9**).

2. Transfer urine into a 50 mL polypropylene plastic tube.

3. Centrifuge 50 mL tube at 10,000 × *g* for 10 min at 4 °C.

4. Aliquot urine supernatant into sterile microtubes (1.7 mL).

5. Store samples at 4 °C and freeze within 2 h of processing at −80 °C for future analysis. Avoid freeze/thaw cycles as this can affect protein stability and degradation.

3.6 Quantification of Secreted MIF in Cell Culture Supernatants, Human Serum, and Urine Using ELISA

1. Before commencing experiments, optimize conditions for running your samples in each commercial ELISA kit that you use.

2. Ensure the concentrations of MIF in your samples lie mid-range within the detection limits of the standard curve for optimal quantification and analysis of MIF (*see* **Note 10**).

3. To quantify concentrations of secreted/circulating MIF follow manufacturer's instructions. Avoid freeze/thaw cycles of your samples as this can affect MIF protein stability and detectability.

3.7 Human Serum/ Plasma MIF Quantification Using a Multiplex Immunoassay

1. Thaw frozen serum/plasma samples at 4 °C

2. Mix samples well by vortexing briefly, then centrifuge for 10 min at 10,000 × *g* at 4 °C. When possible, avoid the use of highly hemolysed, icteric, or lipemic samples as these can invalidate results.

3.7.1 Using Quantibody® Technology

1. For sample preparation and quantification of cytokines, follow manufacturer's instructions. Some optimizations have been developed, as outlined in following steps (*see* **Note 11**).

2. Scan slides using a fluorescence laser scanner. Row signal intensity of protein from unknown samples are read against a 8-point standard curve using the RayBiotech Q-Analyzer software.

3.7.2 Using Luminex™ Technology (Magnetic Beads Only)

1. For sample preparation and quantification of cytokines, follow manufacturer's instructions. We have optimized the following protocol for biosample preparation (*see* **Note 12**):

2. Using a nonbinding plate, prepare samples ready to be transferred onto the Luminex™ kit plate.

3. Make a 1/2 dilution of samples in the transfer plate using 1× Universal Assay Buffer. Change tip after mixing each dilution.

4. Seal the plate and keep at 4 °C.

5. Prepare the standard: centrifuge lyophilized Standard for 10 s at 2,000 × *g*. Reconstitute Standard using 1× Universal Assay Buffer and vortex vial for 10 s. Centrifuge for 10 s at 2,000 × *g*. Incubate on ice for 15 min. Vortex briefly and again centrifuge for 10 s at 2,000 × *g*. Once reconstituted, perform four-fold serial dilutions of standard (change tip after mixing each dilution). Make 1/2 dilutions for standard (as for biosamples) and mix on the transfer plate or in PCR 8-tube strips provided in the Basic kit. The diluted standard is then immediately added to the plate with beads.

6. Bead preparation: prepare the beads during Standard incubation time. Vortex for **30 s** before to prevent aggregation (*see* **Note 13**). Prepare 1× Beads using 1× Wash Buffer for dilution. Vortex. Add additional 200 µL of beads to the reservoir to make sure you have enough.

7. Add 50 µL of beads to each well of the plate. Incubate for at least 2 min on the magnetic plate holder at room temperature.

8. With the plate securely attached to the magnetic plate holder, flick to remove liquid.

9. Tap the plate 1× on clean paper towel.

10. Washing step: dilute Wash Buffer in distilled water to make 1× Wash Buffer. Add 1× Wash Buffer for 30 s to the plate on the magnetic plate holder. With the plate securely attached to the magnetic plate holder, flick to remove liquid (2–3 times). Tap the plate 1× on clean paper towel. Disassemble the plate from the magnetic plate holder.

11. Add diluted Standard and biosamples to the plate. Do not mix.

12. Seal the plate. Incubate on plate mixer (500 rpm) for 2–3 h at room temperature if performing the entire assay on same day. Overnight incubation a 4 °C could be performed if higher assay sensitivity is required (*see* **Note 14**).

13. Wash plate twice as in **step 10** above.

14. Dilute concentrated Detection antibody using Detection Antibody Diluent. Add 1× Detection antibody to the plate.

15. Seal the plate. Incubate on plate mixer (500 rpm) for 30 min at room temperature.

16. Put plate on magnetic plate holder for 2 min to settle down the beads.

17. Wash plate twice as in **step 10** above.

18. Add Streptavidin-PE.

19. Seal the plate. Incubate on plate mixer (500 rpm) for 30 min at room temperature.

20. Wash plate twice as in **step 10** above.

21. Add Reading Buffer to wells.

22. Seal the plate. Incubate on plate mixer (500 rpm) for 5 min at room temperature.

23. Read the plate using a compatible multiplex immunoassay platform, such as MAGPIX®, Luminex™ 200/100, or FLEX-MAP 3D®, depending on the plexing density required.

4   Notes

1. In our laboratory, we routinely use blood serum separation tubes with gel separator, and EDTA blood collection tubes for plasma collection and separation. The use of these tubes results in reliable separation of both serum and plasma, stability of analytes with a short processing time.

2. Other substrates for HRP can also be used, including ABTS (2,2′-Azinobis [3-ethylbenzothiazoline-6-sulfonic acid]-diammonium salt) and OPD (o-phenylenediamine dihydrochloride).

3. Check the manufacturer's instructions supplied with the ELISA kit for concentration of stop solution.

4. In our LPS-induced endotoxemia in vivo models, we routinely use lyophilized *Escherichia coli* 055:B5 strain at concentration of 2 mg/kg. However, this strain has also been shown to also stimulate TLR2-mediated responses. Therefore, if you specifically want to target TLR4 -driven inflammatory responses, LPS-EB VacciGrade™ has been shown to only elicit TLR4-mediated responses.

5. Monitor health status of mice according to institutional guidelines (refer **Table 1**). If mice exhibit severe signs of sepsis and the experiment must be stopped prematurely, or if the mouse dies prematurely, cull the mouse and collect blood. Note that

Table 1
Clinical signs and severity scoring for animal health status monitoring

Signs	Severity score 0	1	2	3
Activity	Normal	Isolated, abnormal posture	Huddled/inactive OR overactive	Moribund or fitting
Alertness/ sleeping	Normal	Dull or depressed	Little response to handling	Unconscious
Breathing	Normal	Rapid, shallow	Rapid, abdominal breathing	Labored, irregular, skin blue
Coat	Normal	Coat rough	Unkempt; wounds, hair thinning	Bleeding or infected wounds, or severe hair loss or self-mutilation
Dehydration	None	Skin less elastic	Skin tenting	Skin tenting and eyes sunken
Eyes	Normal	Wetness or dullness	Discharge	Eyelids matted
Movement/ gait	Normal	Slight incoordination OR abnormal gait	Uncoordinated OR walking on tiptoe OR reluctance to move	Staggering OR limb dragging OR paralysis
Nose	Normal	Wetness	Discharge	Coagulated
Vocalization	Normal	Squeaks when palpated	Struggles and squeaks loudly when handled/palpated	Abnormal vocalization

this table provides guidelines for monitoring over short experimental duration (up to 3–4 h). If you are conducting overnight experiments, you will need to perform more comprehensive monitoring in terms of drinking, eating, body temperature, and urine and fecal outputs.

6. It is important to avoid freeze/thaw cycles of samples as this is detrimental to many serum components.

7. In a recent study published by our laboratory, we showed that unseparated serum should not be stored longer than 3 days at 4 °C to ensure protein stability and prevent degradation [10].

8. There is no clear consensus about long-term storage duration of serum/plasma samples; however, it is generally advised that one should minimize freeze/thaw cycles to less than 3 and make several small aliquots of serum/plasma samples instead of one larger one [11]. It has also been described that cytokines and serum proteins were not shown to degrade and were suitable for analysis if serum/plasma samples had been stored at −80 °C for up to 2 years [12].

9. First morning urine provides the least variability in protein concentration, and mid-stream collection of urine minimizes problems with bacterial contamination [13]. Best practice is to process samples fresh within 30 min of collection.

10. It is recommended that you run samples neat, as well as perform a 1:2 serial dilution of both control and treated samples to ascertain best dilution for use in your experiments.

11. All incubation and washing steps are performed at room temperature with a gentle shaking, except where indicated otherwise. Reconstitute the protein standard mix and prepare a $1/3$ serial dilution using the kit Sample Diluent. Dilute serum/plasma samples $1/2$ using the kit Sample Diluent. Blocking should be done for at least 1 h (original protocol: 30 min) using the kit Sample Diluent. Add 100 µL of protein standards, diluted serum/plasma samples to the slide wells for overnight incubation at 4 °C (original protocol: 1–2 h at room temperature). After the washing step, the reconstituted detection antibody cocktail is added for 1–2 h. Following washing steps postdetection, add the reconstituted streptavidin-conjugated fluorochrome for 1 h. Avoid light exposure. All subsequent steps of the protocol should be performed in the dark with minimal exposure to light. After washing, remove the slide from the chamber and gasket, and place into 4-slide slots in a kit-supplied 50 mL centrifuge tube and fill with wash buffer. After the last washing step, carefully dry slides by centrifuging at $1,000 \times g$ for 3 min. An extra drying step has been added to the original protocol so the slides can be air-dried overnight in a clean 4-slide holder tube with no lid on. Important to note, slide wells should never dry out for prolonged period of time.

12. This protocol is meant to be used with ProcartaPlex^{TM} immunoassay. The entire assay procedure is generally completed within a day; however, it can be split in 2 days for better assay sensitivity. For all incubation steps, plate (beads) should be protected from light exposure, being covered with black lid. Always use new plate seal film for each step. As a general note, when adding reagent, it is recommended to touch the side of the well (wall) with the extremity of the tip and to change tip each time, with exception of the washing step.

13. Never centrifuge beads.

14. For overnight incubation, shake the plate for only 30 min to 1 hour on plate mixer (500 rpm) at room temperature. Transfer the plate at 4 °C for overnight incubation (no incubation on

plate mixer is required). The plate must be kept in a Styrofoam box with moist paper towel on level surface. After overnight incubation, shake the plate for an additional 30 min (500 rpm) at room temperature.

References

1. Lang T et al (2015) MIF: implications in the pathoetiology of systemic lupus erythematosus. Front Immunol 6:577

2. Bloom BR, Bennett B (1966) Mechanism of a reaction in vitro associated with delayed-type hypersensitivity. Science 153(3731):80–82

3. David JR (1966) Delayed hypersensitivity in vitro: its mediation by cell-free substances formed by lymphoid cell-antigen interaction. Proc Natl Acad Sci U S A 56(1):72–77

4. Baugh JA, Bucala R (2002) Macrophage migration inhibitory factor. Crit Care Med 30 (1 Supp):S27–S35

5. Lan HY (2008) Role of macrophage migration inhibition factor in kidney disease. Nephron Exp Nephrol 109(3):e79–e83

6. Calandra T, Roger T (2003) Macrophage migration inhibitory factor: a regulator of innate immunity. Nat Rev Immunol 3 (10):791–800

7. Vincent FB et al (2019) Serum and urinary macrophage migration inhibitory factor (MIF) in primary Sjogren's syndrome. Joint Bone Spine 86(3):393–395

8. Vincent FB et al (2018) Analysis of serum macrophage migration inhibitory factor and D-dopachrome tautomerase in systemic sclerosis. Clin Transl Immunol 7(12):e1042

9. Vincent FB et al (2018) Analysis of urinary macrophage migration inhibitory factor in systemic lupus erythematosus. Lupus Sci Med 5 (1):e000277

10. Vincent FB et al (2019) Effect of storage duration on cytokine stability in human serum and plasma. Cytokine 113:453–457

11. Tuck MK et al (2009) Standard operating procedures for serum and plasma collection: early detection research network consensus statement standard operating procedure integration working group. J Proteome Res 8(1):113–117

12. Keustermans GC et al (2013) Cytokine assays: an assessment of the preparation and treatment of blood and tissue samples. Methods 61 (1):10–17

13. Thomas CE et al (2010) Urine collection and processing for protein biomarker discovery and quantification. Cancer Epidemiol Biomark Prev 19(4):953–959

Chapter 6

Flow Cytometry Phenotyping of Bone Marrow-Derived Macrophages from Wild-Type and *Mif*^{−/−} Mice

Jacqueline K. Flynn, Nadia S. Deen, and James Harris

Abstract

Phenotyping cells by flow cytometry is a powerful way to identify cell type and any morphological changes during cell culture. The staining procedure used in this chapter enables the characterization of mouse macrophages by a flow cytometry antibody panel which can be used for both bone marrow-derived macrophages (BMM) and macrophages derived from other tissues, such as the mouse spleen or peritoneal cavity. The surface and intracellular staining methods are versatile and can be applied to flow cytometry staining of several different cell types by changing the surface markers used with knowledge of which receptors are expressed on different cell types.

Key words BMM, Macrophages, Fluorochromes, Flow cytometry, Cell surface receptors, MIF

1 Introduction

Macrophages are a key innate immune cell type which constantly survey their environment for tissue damage and invading pathogens. They are antigen-presenting cells, able to phagocytose foreign antigens and present them to T cell and B cells, maintain the health of our tissues by removing dead and dying cells, and respond to danger signals via their surface receptors [1]. Macrophages are derived from the myeloid lineage and are released into circulation as bone marrow-derived precursors, monocytes [1].

Bone marrow-derived macrophages (BMM) can be produced by isolating their hematopoietic precursors from the leg bones of mice and culturing them in the presence of macrophage colony-stimulating factor (M-CSF) or media with equivalent growth factors and cytokines [2]. This technique gives rise to F4/80⁺ (macrophage marker) CD11b⁺ (myeloid marker) double positive macrophages which are easily identified by flow cytometry and widely used in research (Fig. 1) [3]. An additional advantage is that these cells can be obtained from genetically modified mice,

James Harris and Eric F. Morand (eds.), *Macrophage Migration Inhibitory Factor: Methods and Protocols*,
Methods in Molecular Biology, vol. 2080, https://doi.org/10.1007/978-1-4939-9936-1_6,
© Springer Science+Business Media, LLC, part of Springer Nature 2020

A) Day 8 BMM from wildtype mice

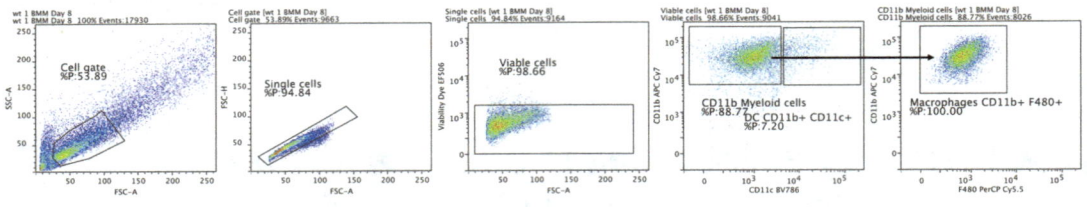

B) Day 8 BMM from Mif-/-

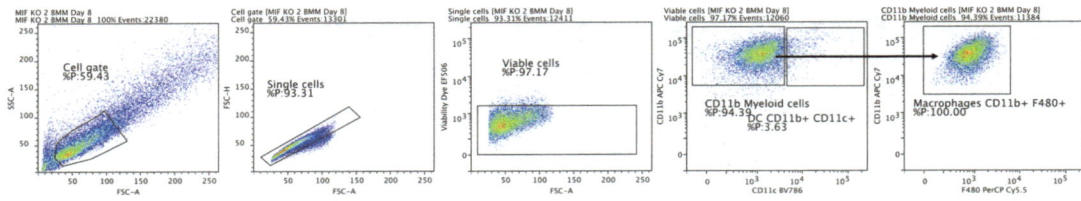

Fig. 1 Phenotyping day 8 BMM from wild-type and $Mif^{-/-}$ mice. To identify macrophages in the BMM cells the cells are first gated using a series of gates to ensure viable single cell populations. Moving from left to right these gates include a general viable cell gate to exclude apoptotic and dying cells (FSC-A versus SSC-A), which also distinguishes cells based on size and granularity, then a single cell gate is used (FSC-A versus FSC-H) to exclude doublets from the analysis. The third gate is the viable cell gate using the fixable Viability Dye EF506, which is shown against FSC-A. The viable cells are able to exclude the dye and are thus negative. The fourth gate and plot shows CD11b + cells to select for myeloid cells. This plot is against CD11c to exclude dendritic cells from the analysis. Next the CD11b+CD11c− cells are used to identify the macrophage population. This is shown in the fifth and final gate where double positive CD11b + F4/80+ macrophages are shown. This gating strategy is used on both (**a**) wild-type and (**b**) $Mif^{-/-}$ day 9 BMM

for example, macrophage inhibitory factor knockout ($Mif^{-/-}$) mice.

MIF is a pro-inflammatory molecule produced by most immune cells, including macrophages [4]. It is induced by inflammatory stimuli, such as lipopolysaccharide (LPS), and cytokines, such as TNF and IFN-γ, and has been linked to a pathogenic role in inflammatory diseases and tumor progression [4]. MIF itself induces the induction of other cytokines, including TNF, IL-6, and IL-1 family members [5–7]. The use of BMM from wild-type and $Mif^{-/-}$ mice provides a great tool with which to study the role of MIF in macrophages and whether they have any morphological or phenotypic changes.

Macrophages can be classified into different subsets based largely upon their immune functions. M1 macrophages, termed classically activated, mediate defenses against viruses, bacteria, and protozoa [8–10]. They also commonly secrete pro-inflammatory cytokines, including IL-1β, IL-12, IL-18, TNF, as well as reactive oxygen species (ROS). M2 macrophages, termed alternatively activated, have a more an anti-inflammatory role, secreting IL-10 and TGF-β, and can mediate wound healing [8, 11]. There is plasticity between macrophage activation states, as these phenotypes can be

Table 1
Phenotypic characteristics of macrophages

Antibody	Macrophages	M1 Macrophages	M2 Macrophages
CD11b	+	+	+
F4/80	+	+	+
CD11c	−	−	−
MHC II	+	+	+
TLR2	+	+	−
TLR4	+	+	−
CD80	+	+	−
CD206	+	−	+

CD11c is a dendritic cell marker used to exclude dendritic cells from the analysis. M2 macrophages include all subsets, M2a, M2b, and M2c

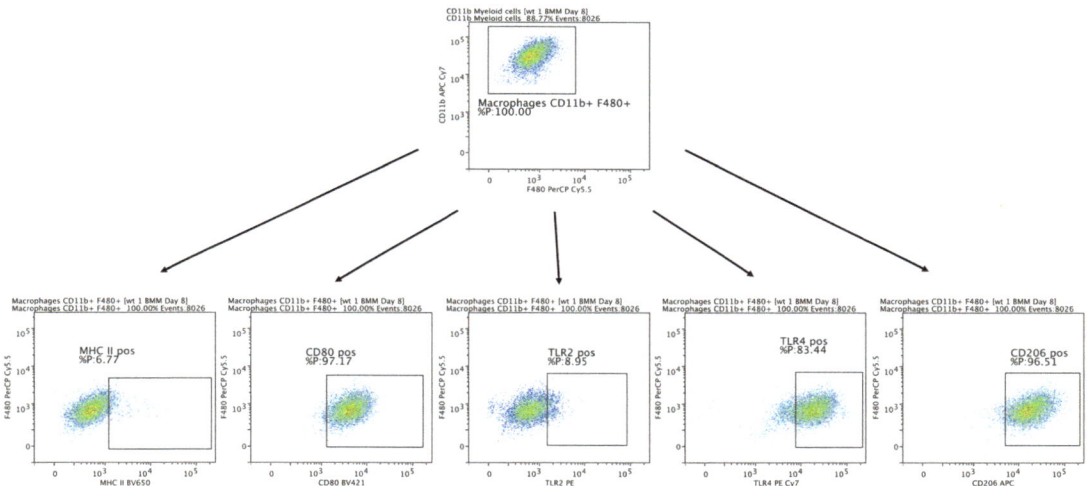

Fig. 2 M1 and M2 phenotyping day 8 BMM from wild-type mice. Proceeding from the gating strategy shown in Fig. 1, the wild-type mouse macrophages (CD11b⁺F4/80⁺) can then be examined for M1 (CD80, TLR2, TLR4) and M2 (CD206) marker expression and positivity. The box gates shown in each plot indicate the percentage of macrophages positive for that marker. CD206 was detected via intracellular staining

reversible under certain conditions [12]. The macrophage phenotype can also be identified by surface markers for M1 and M2 populations (Table 1) with M1 macrophages upregulating CD80, TLR2, and TLR4 and M2 macrophages upregulating CD206 (Mannose receptor) (Figs. 2, 3, and 4) [12].

This chapter focuses on cell surface and intracellular staining to identify BMM from wild-type and *Mif*⁻/⁻ mice and phenotype

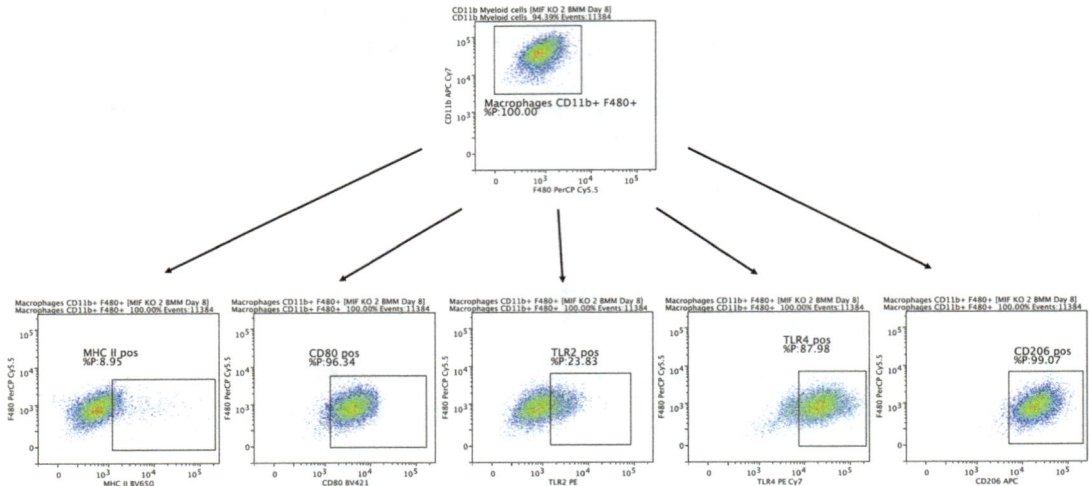

Fig. 3 M1 and M2 phenotyping day 8 BMM from *Mif*⁻/⁻ mice. Proceeding from the gating strategy shown in Fig. 1, the MIF KO mouse macrophages (CD11b⁺F4/80⁺) can then be examined for M1 (CD80, TLR2, TLR4) and M2 (CD206) marker expression and positivity. The box gates shown in each plot indicate the percentage of macrophages positive for that marker. CD206 was detected via intracellular staining

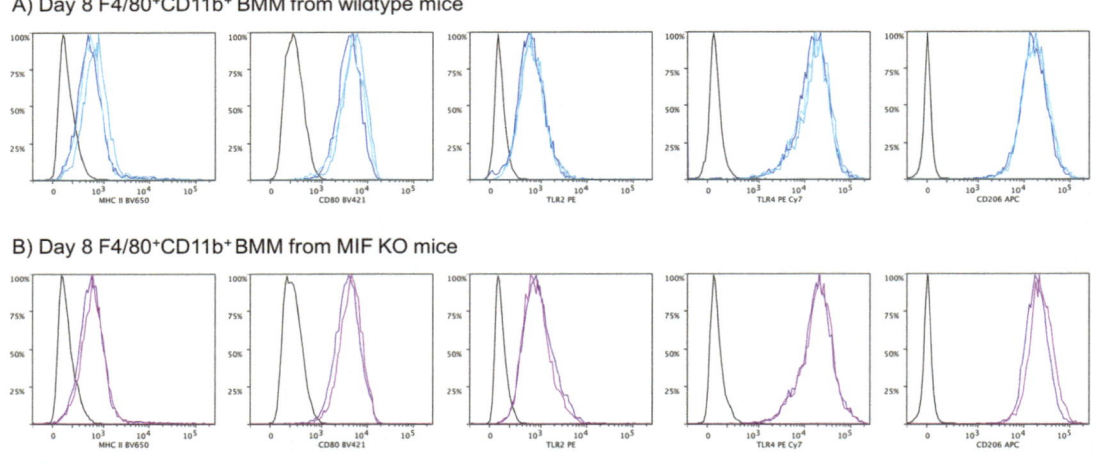

Fig. 4 Histogram analysis of M1 and M2 phenotyping of day 8 BMM from wild-type and *Mif*⁻/⁻ mice. Proceeding from the gating strategy shown in Fig. 1, both the wild-type and MIF KO mouse macrophages (CD11b⁺F4/80⁺) can then be examined for M1 (CD80, TLR2, TLR4) and M2 (CD206) marker expression via histograms. The receptor expression is displayed as histograms to show the spread of each receptor expression for (**a**) wild-type in blue (*n* = 3) and (**b**) *Mif*⁻/⁻ in purple (*n* = 2). CD206 was detected via intracellular staining

them using a panel of M1 and M2 markers. As this technique incorporates intracellular staining, this protocol can be easily adapted at the intracellular staining step by adding to the panel relevant cytokine antibodies, such as IL-6 and TNF, to determine a functional profile of the macrophages.

2   Materials

2.1   Macrophage Polarization

1. Lipopolysaccharide (LPS): final concentration 100 ng/mL, stock: 1 mg/mL in sterile milli-Q H_2O.

2. Recombinant murine IFNγ: final concentration 50 ng/mL, stock: 50–100 μg/mL in sterile milli-Q H_2O with 0.1% bovine serum albumin (BSA).

3. Recombinant murine IL-4: final concentration 10 ng/mL, stock: 10–100 μg/mL in sterile milli-Q H_2O with 0.1% BSA.

4. Recombinant murine IL-10: final concentration 10 ng/mL, stock: 10–100 μg/mL in sterile milli-Q H_2O with 0.1% BSA.

5. Recombinant murine IL-13: final concentration 10 ng/mL, stock: 10–100 μg/mL in sterile milli-Q H_2O with 0.1% BSA.

2.2   Intracellular Flow Cytometry Staining

1. 5 mL polypropylene flow cytometry tubes capped (*see* **Note 1**).

2. FACS Buffer; 500 mL sterile PBS (Ca/Mg^{2+} free), 0.1% BSA, 1 mM EDTA (*see* **Note 2**). Filter sterilize any unsterile components to keep buffer sterile. Store at 4 °C.

3. Fixation Buffer; 1% paraformaldehyde in PBS (Ca/Mg^{2+} free) (*see* **Note 3**). Store at 4 °C.

4. Permeabilization Buffer: 0.1% Saponin in PBS (Ca/Mg^{2+} free) (*see* **Note 4**). Store at 4 °C.

5. Flow Cytometry Antibodies; example macrophage Panel; CD11b APC Cy7, CD11c BV786, F480 PerCP Cy5.5, MHC II BV650, CD80 BV421, TLR2 PE, TLR4 PE Cy7, CD206 APC (Tables 1 and 2).

6. Fixable Viability dye (for example, EF506 to fit in the macrophage panel above, Table 3, *see* **Note 5**).

7. Compensation Beads. Store at 4 °C.

3   Methods

3.1   Macrophage Polarization

1. Culture and harvest bone marrow-derived macrophages (BMM) as described in Chapter 2.

2. On day 6–8, treat BMM with activators of polarization: For M1; 100 ng/mL LPS or 100 ng/mL LPS + 50 ng/mL IFNγ for 12–24 h, for M2; 10 ng/mL IL-4 and/or 10 ng/mL IL-10 or 10 ng/mL IL-13 for 12–24 h.

3.2   Intracellular Flow Cytometry Staining

All procedures should be carried out in a class II biological hazard cabinet to maintain a sterile environment. A brief guide on the use of viability markers, compensation controls, experimental controls,

Table 2
Flow cytometry macrophage panel antibody amounts

Flow antibody	Fluorochrome	Amount per compensation tube	Amount per cell sample (staining in 50 μL)	Amount per ten sample Master Mix (500 μL total volume, 50 μL per sample)
CD11b	APC-Cy7	1 μL	0.25 μL	2.5 μL
F4/80	PerCP Cy5.5	1 μL	0.25 μL	2.5 μL
CD11c	BV786	1 μL	0.25 μL	2.5 μL
MHC II	BV650	1 μL	0.25 μL	2.5 μL
Viability Dye	EF506	N/A	0.25 μL	2.5 μL
TLR2	PE	1 μL	0.25 μL	2.5 μL
TLR4	PE-Cy7	1 μL	0.25 μL	2.5 μL
CD80	BV421	1 μL	0.25 μL	2.5 μL
CD206	APC	1 μL	0.25 μL	2.5 μL

Please note depending upon supplier of flow antibody you use, you may need to titrate the amount of antibody for your experiments. These volumes are for using a total volume of 50ul per sample for staining and using the antibodies at a 1 in 200 dilution

Table 3
Flow cytometry macrophage antibodies and Fortessa X20 instrument configuration

Flow antibody	Fluorochrome	Fortessa laser	Description
CD80	BV421	405 nm	V 450/50
Viability dye	EF506	405 nm	V 525/50
MHC II	BV650	405 nm	V 670/30
CD11c	BV786	405 nm	V 780/60
F4/80	PerCP Cy5.5	488 nm	B 710/50
TLR2	PE	561 nm	YG 586/15
TLR4	PE Cy7	561 nm	YG 780/60
CD206	APC	640 nm	R 670/30
CD11b	APC Cy7	640 nm	R 780/60

Please note refer to the spectral guides and instrument configuration to ensure these antibodies can be used on your instrument

panel design, and the titration of flow antibodies have been added to the notes section (*see* **Notes 5–9**).

For each experiment, you need to have single color compensation controls, unstained cell alone tubes, and sample tubes.

3.2.1 Prepare Single-
Color Compensation Tubes

1. Prepare and label flow tubes for each antibody/fluorophore you are testing (Table 2).

2. Vortex compensation beads well.

3. Add one drop of the compensation bead solution to each tube (*see* **Note 6**).

4. Add 1 μL of each antibody to the bottom of each compensation tube, into the bead solution and pipette up and down well (Table 2).

5. Stain the beads for 30 min at 4 °C in the dark.

6. Wash the beads by adding 1 mL of FACS buffer to each tube and spin cells at $300 \times g$ for 5 min.

7. Remove supernatant and add 200 μL of FACS FIX.

8. Fix the beads for 30 min at 4 °C in the dark.

9. Wash the beads by adding 1 mL of FACS buffer to each compensation tube and spin cells at $300 \times g$ for 5 min.

10. Remove supernatant carefully and resuspend the beads in 100 μL of FACS buffer.

11. Store flow tubes at 4 °C covered in foil. The compensation tubes can be stored, once fixed and washed, for up to 1 week.

3.2.2 Prepare Unstained
Cells Alone Tubes
and Sample Tubes

1. Harvest BMM cells from the wild-type and $Mif^{-/-}$ day 8 BMM cultures and perform a cell count. Centrifuge at $300 \times g$ for 5 min and resuspend a total of 5×10^5 in 200 μL FACS buffer.

2. Aliquot 100 μL of cell suspension into a flow cytometry tube labeled as *cells alone* (unstained) with mouse type—wild type or $Mif^{-/-}$. These cells will allow correct setup of voltages on the flow cytometer. *Cells alone* tubes are treated the same as the *sample* tubes, with exception that the cells will remain unstained throughout the subsequent steps below (*see* **Note 8**).

3. The remaining 100 μL of cell suspension will be used to phenotype your *sample* tubes, the BMM cells using intracellular staining methods. Label each tube with macrophage panel, mouse type (wild type or $Mif^{-/-}$), and other identifying labels as required.

4. Centrifuge both the *cells alone* tubes and *samples* tubes at $300 \times g$ for 5 min.

5. Carefully remove the supernatant.

6. To the *cells alone* tubes, add 50 μL of FACS buffer to the tubes and resuspend well. Incubate cells for 30 min at 4 °C.

7. To the *sample* tubes, stain cells with surface marker antibodies (Table 2). Make up a master mix of flow cytometry antibodies (for example, this macrophage panel uses the antibodies at a

1/200 dilution. Ensure all antibodies are titrated prior to use *see* **Note 9**). Add 50 μL of antibody master mix per tube and resuspend the cells well in the antibody master mix. Stain cells for 30 min at 4 °C in the dark (in the fridge, covered in foil, *see* **Note 10**).

8. Add 1 mL of FACS buffer to each of the cells alone and sample tubes and spin cells at $300 \times g$ for 5 min.

9. Remove the supernatant from all tubes and resuspend the cells in 100 μl of fixation buffer. Fix the *cells alone* and *sample* tubes for 20 min at 4 °C (in the fridge, covered in foil).

10. Resuspend the cells from all tubes in 1 mL FACS buffer and centrifuge at $300 \times g$ for 5 min.

11. Remove the supernatant and resuspend the cells from all tubes in 100 μL permeabilization buffer and incubate for 20 min at 4 °C (in the fridge, covered in foil).

12. Add 200 μL of permeabilization buffer to all tubes and centrifuge at $300 \times g$ for 5 min.

13. For the *cells alone* tubes, resuspend the cells in 50 μL permeabilization buffer and incubate at 4 °C for 30 min (in the fridge).

14. For the *sample* tubes, create a master mix of flow cytometry antibodies for intracellular staining (CD206 antibody, used at a 1/200 dilution) in permeabilization buffer (50 μL buffer per sample) (Table 2). Remove the supernatant from the washed cells and resuspend in 50 μL of intracellular flow antibody master mix (*see* **Note 11**). Incubate at 4 °C for 30 min (in the fridge, covered in foil).

15. Add 200 μL of permeabilization buffer to each tube and centrifuge at $300 \times g$ for 5 min.

16. Remove supernatant carefully and resuspend in 200 μL of FACS buffer (*see* **Notes 12** and **13**).

17. Store flow tubes at 4 °C covered in foil (in the fridge) for up to 1 week.

18. Run all flow tubes within a week on an LSR cytometer or similar instrument with a configuration to allow detection of the panel of antibodies (Table 3).

4   Notes

1. It is critical that the flow cytometry tubes used make a proper seal on your flow cytometer.

2. Instead of BSA and EDTA you can use 2% FCS in FACS buffer.

3. A fixative of 1–4% paraformaldehyde will work on most cell types. Please check your institution's requirements for any infectious cell work, as often a 4% fixative is required.

4. Permeabilization buffers commonly contain either 0.1% saponin or 0.1% Triton X-100 in PBS. Commercial buffers are also available, some of which are a combined fixation and permeabilization buffer. If you are using a combination buffer omit the fixation step from the protocol (Subheading 3.2.2, **steps 9–11** and **22**, and fix and permeabilize your cells at the same time).

5. Include a good viability marker to allow the easy removal of dead cells from your analysis. Dead and dying cells may give false positives from their autofluorescent properties and cause increased nonspecific antibody binding.

6. Good compensation controls are vital. This is due to spectral overlap of each fluorochrome's emission spectra which can result in detection in another or several channels. The use of compensation beads can assist compensation, especially when with compensating a rare surface marker or a dim dye on primary cells.

7. Additional controls such as fluorescence minus one controls (FMO) will provide a measure of spillover in a given channel by showing the spread of all other antibodies into the unlabeled channel. Isotype controls are also good for determining nonspecific antibody binding.

8. Cells can be stained in 96-well v-bottom plates or tubes. For staining in 96-well plates adjust wash step volumes accordingly to have a maximum of 200 μL per well.

9. Antibody titration is critical. Too much antibody can move both your negative and positive cell populations along the axis and too little antibody can make it very difficult to determine negative and positive populations. Too much and too little antibody can create difficulties for compensation and gating negative and positive populations.

10. For good panel design knowing the density of surface molecule you are detecting on each cell type, the brightness of fluorochrome, and the spread of each fluorochrome will assist greatly in your choice of antibody and data analysis.

11. If having trouble getting good surface receptor staining for some of your antibodies, you can also try staining at room temperature covered in foil.

12. Cell suspension volume will depend upon cell number and type of flow cytometer used for analysis.

13. Some fluorochromes, especially tandem dyes need to be used with caution as they can be prone to uncoupling, and sensitive to photobleaching and extended incubation in *fixation* buffers. In this panel PE-Cy7 and APC-Cy7 can react with fixative and are more sensitive to light causing a loss of signal; thus, washing off the fixative and keeping samples in the dark is required. Use of an H7 rather than Cy7 conjugate can improve stability.

References

1. Murray PJ, Wynn TA (2011) Protective and pathogenic functions of macrophage subsets. Nat Rev Immunol 11:723–737

2. Assouvie A, Daley-Bauer LP, Rousselet G (2018) In: Rousselet G (ed) Growing murine bone marrow-derived macrophages, Macrophages: methods and protocols in molecular biology, vol 1784. Springer, New York

3. Devaud C et al (2014) Tissues in different anatomical sites can sculpt and vary the tumor microenvironment to affect responses to therapy. Mol Ther 22(1):18–27

4. Lee JP et al (2016) Loss of autophagy enhances MIF/macrophage migration inhibitory factor release by macrophages. Autophagy 12 (6):907–916

5. Lang T et al (2018) Macrophage migration inhibitory factor is required for NLRP3 inflammasome activation. Nat Commun 9(1):2223

6. Toh M-L et al (2006) Regulation of IL-1 and TNF receptor expression and function by endogenous macrophage migration inhibitory factor. J Immunol 177(7):4818

7. Chuang C-C et al (2010) Macrophage migration inhibitory factor regulates interleukin-6 production by facilitating nuclear factor-kappa B activation during Vibrio vulnificus infection. BMC Immunol 11:50–50

8. Mills CD et al (2000) M-1/M-2 macrophages and the Th1/Th2 Paradigm. J Immunol 164 (12):6166–6173

9. Classen A, Lloberas J, Celada A (2009) Macrophage activation: classical vs. alternative. Methods Mol Biol 531:29–43

10. Flynn JK, Gorry PR (2015) Role of macrophages in the immunopathogenesis of HIV-1 infection. In: Shapshak JTSP, Somboonwit C, Kuhn JH (eds) Global virology i-identifying and investigating viral diseases. Springer, New York

11. Herbein G, Varin A (2010) The macrophage in HIV-1 infection: from activation to deactivation? Retrovirology 7:33

12. Roszer T (2015) Understanding the mysterious M2 macrophage through activation markers and effector mechanisms. Mediat Inflamm 2015:16

Chapter 7

Genotyping Two Promoter Polymorphisms in the MIF Gene: A −794 CATT$_{5-8}$ Microsatellite Repeat and a −173 G/C SNP

Lin Leng, Edwin Siu, and Richard Bucala

Abstract

Macrophage migration inhibitory factor (MIF) is an upstream proinflammatory cytokine encoded by a functionally polymorphic locus. The promoter region of the human *MIF* gene contains two polymorphisms. A variable nucleotide tandem repeat at position −794 comprises five to eight CATT repeats (referred to henceforth by numbers from 5 to 8, rs5844572). Gene reporter assays show a proportional increase in transcription with CATT repeat number; the 5-repeat allele leads to low expression, and the 6-, 7-, and 8-repeat alleles lead to correspondingly higher expression of MIF. A second *MIF* promoter polymorphism comprises a G-to-C single nucleotide polymorphism (SNP) at position −173 (rs755622), which is in strong linkage disequilibrium with −794 7-CATT and is associated with arthritis clinical severity and higher serum and synovial fluid MIF levels. This allele also has been reported to confer improved survival in patients with outpatient pneumonia. In this chapter, we will introduce the methods of genotyping CATT$_{5-8}$ repeats and the MIF −173 G/C from human samples.

Key words MIF polymorphisms, CATT repeats, −173SNP, Capillary electrophoresis, Real-time PCR

1 Introduction

Macrophage migration inhibitory factor (MIF) is an upstream immunoregulatory cytokine that contributes to the pathogenesis of autoimmunity, infectious diseases, and cancer [1–3]. MIF counter-regulates the immunosuppressive action of glucocorticoids and promotes the survival of different cell types by inhibiting activation-induced apoptosis [4, 5]. In the case of macrophages, autocrine/paracrine MIF release sustains high expression levels of microbial pattern recognition receptors, innate cytokines, and prostaglandins [5, 6]. There is a single *MIF* gene (NCBI-Entrez ID: 4282, Ensembl ID: ENSG00000240972) in the human genome (chromosomal location: 22q 11.23). Human genetic studies have identified two distinct polymorphisms in the MIF promoter: (A) a 4-nucleotide microsatellite (CATT) in the *MIF* promoter that is present in 5 to 8 copies (CATT$_{5-8}$, *rs5844572*) at position −794

James Harris and Eric F. Morand (eds.), *Macrophage Migration Inhibitory Factor: Methods and Protocols*,
Methods in Molecular Biology, vol. 2080, https://doi.org/10.1007/978-1-4939-9936-1_7,
© Springer Science+Business Media, LLC, part of Springer Nature 2020

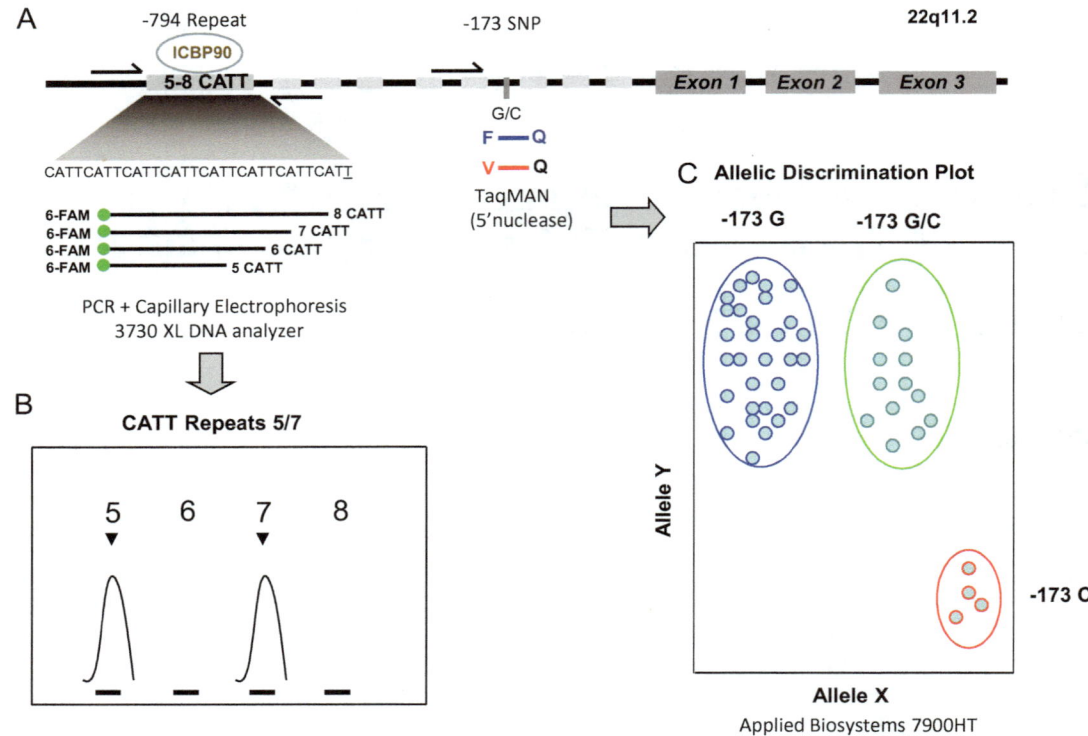

Fig. 1 Genotyping of MIF Alleles. (**a**) Polymorphic structure of the human MIF gene (*MIF*) show the −794 CATT repeat, and the −173 G/C single nucleotide polymorphism, (**b**) microsatellite fragment analysis by capillary electrophoresis peaks for a sample with CATT repeats 5/7, (**c**) TaqMan allelic discrimination plot with results for multiple samples, showing the GG, GC, and CC SNPs

(Fig. 1) [7, 8] and (B) a G/C single nucleotide polymorphism (SNP, *rs755622*) at position −173 (Fig. 1) [9]. Functional studies indicate that CATT repeat number is associated with inducible *MIF* expression such that the $CATT_5$ repeat is a low expression allele and the $CATT_6$, $CATT_7$, and $CATT_8$ repeats are progressively higher expression alleles [7]. These promoter variants occur commonly in the population (minor allele frequency > 5%), and higher CATT repeat number is linked to the susceptibility or the clinical severity of autoimmune inflammatory disease [10–12]. Infectious, oncogenic, and neurodevelopmental disorders with an inflammatory pathogenesis also have been associated with the *MIF* microsatellite repeats, with odds ratios as high as 2.7 for meningococcal sepsis and 9.7 for metastatic prostate cancer [13–15]. The −173C SNP allele, which is in linkage disequilibrium with CATT, is associated with earlier disease onset and severity in diverse disorders including inflammatory polyarthritis, pediatric nephrotic syndrome, alopecia areata, and inflammatory bowel disease. Furthermore, the presence of the −173C SNP allele has been shown to associate with steroid resistance in a range of diseases including colitis, juvenile arthritis, and nephrotic syndrome [16–20]. Therefore, the −173C SNP allele in the *MIF* promoter is a candidate marker of steroid requirements and disease activity in autoimmune disease. Emerging knowledge about the association of the variant alleles of different genes with the human diseases have focused attention on the development of technologies to rapidly discriminate between different genotypes. The genotyping methods are important both to facilitate the acquisition of genotyping data and to better translate this knowledge into diagnostic and therapeutic applications [21].

To genotype the CATT repeats found at the MIF −794 locus (*rs5844572*), we use PCR amplification with fluorescently labeled primers for the target region followed by microsatellite fragment length analysis via capillary electrophoresis as described in Subheadings 3.1 and 3.2. The different repeat numbers produce fragments that migrate at different rates in capillary electrophoresis and the fluorescent label incorporated in the PCR product allows detection and sizing of the microsatellite length. We perform this assay on the ABI 3730xl Genetic Analyzer which accepts a 96-well plate format followed by analysis in GeneMapper software.

Here, we present two protocols for genotyping the G/C SNP found at the MIF −173 locus (*rs755622*). The TaqMan PCR protocol described in Subheadings 3.3–3.6 is our primary approach and provides automated allelic discrimination. The assay is performed in the 384-well format on the ABI 7900HT Real-Time PCR and analysis is done in SDS software allowing high-throughput processing. Each TaqMan probe is specific to one allele and is labeled with a separate fluorescent dye, FAM for the major allele (G), and VIC for the minor allele (C) (Fig. 2). In some cases, the primary genotyping method with the TaqMan kit is not successful. Alternatively, the polymerase chain reaction-restriction fragment length polymorphism (PCR-RFLP) analysis method described in Subheading 3.7 may work in the subset of samples that cannot be successfully genotyped with the TaqMan kit. This method also allows for the direct visualization of the MIF -173 locus genotype on an agarose gel [22, 23]. It is more labor intensive than TaqMan but has the advantage of being relatively robust, cheap, and may be useful to labs that lack access to the specialized real-time PCR equipment and software necessary for implementing the TaqMan assay.

2 Materials

2.1 *Purified gDNA*

The quality, accuracy, and amplifiability of DNA can be significantly affected by characteristics of the sample itself, the collection method, and the purification process. The main sources we use for samples of human DNA are blood (*see* **Note 1**) [24, 25], saliva,

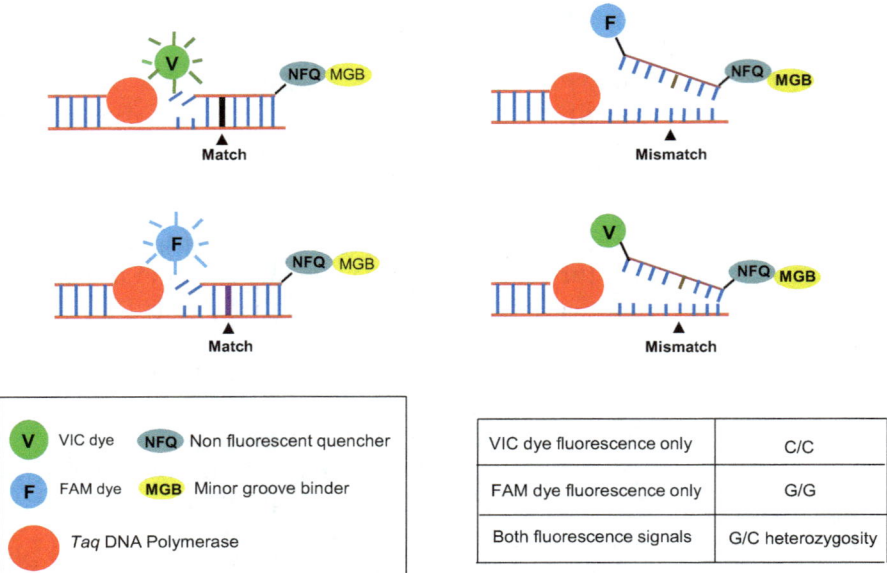

Fig. 2 Genotyping the MIF −173 G/C SNP by TaqMan Allelic Discrimination. The TaqMan probe will bind only to a specific sequence that is present at MIF −173 locus. The intact probe does not normally have fluorescent signal due to the presence of the nonfluorescent quencher. The fluorescent signal is only activated when the *Taq* DNA polymerase encounters the DNA template-bound TaqMan probe and frees the conjugated fluorescent dye molecule. The corresponding fluorescence signal will indicate the single nucleotide polymorphisms present in the MIF promoter region

cells, and tissues. We primarily use the Easy-DNA gDNA Purification Kit from Invitrogen to isolate and purify gDNA from whole blood, buffy coat or plasma. Other commercial kits are available, such as the ORAgene-DNA kit from DNA Genotek, which we have used to isolate gDNA from saliva [26] and the QIAamp DNA Mini Kit from QIAGEN for isolating gDNA from dried blood spot samples on FTA Micro Cards. A distinct advantage of both the FTA Micro Cards and the Oragene-DNA kit is that they permit collection of samples that are stable for long-term storage at room temperature.

Because both the quantity and quality of the DNA are paramount to genotyping success, we recommend that all samples are initially assayed on a nano-scale spectrophotometer, which can allow accurate determination of DNA quantity and quality using only 1–2 μL of your purified gDNA sample. Ideally, your 260/280 ratio should be greater than or equal to 1.8 and the 260/230 ratio 2.0–2.2. The optimal human gDNA concentration is 10 ng/μL for our PCR reactions but we have had success with concentrations ranging from 2.5 to 100 ng/μL (*see* **Note 2**) [27, 28].

Once your samples have been purified, all gDNA should be stored at −20 °C to 4 °C until you are ready to perform the PCR.

2.2 Genotyping the MIF − 794 CATT₅₋₈ Polymorphism by Microsatellite Fragment Length Assay

1. PCR SuperMix, Invitrogen (*see* **Note 3**).
2. Primers, diluted to 10 µM working concentration
   CATT forward: 5′–TGC AGG AAC CAA TAC CCA TAGG–3′
   CATT reverse: 5′–XAAT GGT AAA CTC GGG GAC–3′ (where X = 6-FAM label) (*see* **Note 4**).
3. Agarose, molecular biology grade.
4. GelRed Nucleic Acid Gel Stain.
5. MicroAmp Optical 96-well Reaction Plate, Applied Biosystems.
6. MicroAmp Optical Adhesive Film, Applied Biosystems.
7. TBE buffer, molecular biology grade.
8. ddH₂O, molecular biology grade,
9. Purified genomic DNA samples for genotyping (5–25 ng/µL).
10. DNA loading buffer, for agarose gel electrophoresis.
11. DNA 100 bp size standard, for agarose gel electrophoresis.
12. Hi-Di Formamide, Applied Biosystems (*see* **Note 4**).
13. GeneScan-500 LIZ Size Standard, Applied Biosystems (*see* **Note 4**).

2.3 Genotyping the MIF − 173 G/C SNP by TaqMan Allelic Discrimination Assay

1. TaqMan Genotyping Master Mix, Applied Biosystems (*see* **Note 5**).
2. 40× Assays-On-Demand SNP Genotyping Assay Mix, Applied Biosystems #4351379, custom design C_2213785_10, rs755622 (*see* **Note 5**)
3. MicroAmp Optical 384-well reaction plate, Applied Biosystems.
4. MicroAmp Optical Adhesive Film, Applied Biosystems.
5. Purified genomic DNA samples for genotyping, 5–25 ng/µL.
6. ddH₂O, molecular biology grade.

2.4 Genotyping the MIF − 173 G/C SNP by PCR-RFLP Assay

1. KAPA Taq HotStart PCR Kit: including MgCl₂, dNTPs, 5× Buffer, Taq Polymerase Enzyme (*see* **Note 6**).
2. ddH₂O.
3. Primers, diluted to 10 µM working concentration.
   MIF-rflp forward: 5′–ACT AAG AAA GAC CCG AGGC–3′
   MIF-rflp reverse: 5′–GGG GCA CGT TGG TGT TTAC–3′
4. *AluI* restriction enzyme and 10× CutSmart buffer, New England Biolabs (*see* **Note 6**).
5. ddH₂O, molecular biology grade.
6. Agarose, molecular biology grade.

7. Low Melting Point Agarose, for making 3% gel.

8. GelRed Nucleic Acid Gel Stain.

9. DNA Loading Buffer, for agarose gels.

10. DNA 100 bp size standard, for agarose gel electrophoresis.

11. Purified genomic DNA samples (5–25 ng/μL).
    This protocol also requires known positive control DNA for GG, CC, and CC alleles.

3   Methods

3.1   Genotyping the MIF − 794 CATT$_{5-8}$ Polymorphism by Microsatellite Fragment Length Assay

1. Thaw, mix well, and centrifuge all reagents and samples before use. Prepare a PCR master mix with the Invitrogen PCR Super-Mix, on ice, per Table 1, without the template DNA. Prepare 15–20% more master mix than needed (for example, when preparing master mix for 10 samples and controls, make enough master mix for 12 reactions). Mix well and spin down in centrifuge. Always include both a positive DNA sample, preferably of a known size, and a negative (water), as a control. The positive control is for troubleshooting PCR amplification and allows you to monitor sizing precision. The negative control is to detect unwanted amplification of contaminating DNA. Amplify at least one control DNA sample in every PCR run.

2. Aliquot 24.5 μL of the master mix into each PCR tube. Add 0.5 μL of genomic DNA into each reaction well, changing tips between every well. For the negative control well, add 0.5 μL of ddH$_2$O instead. The final reaction volume will be 25 μL. Check to make sure caps are on securely, then mix and centrifuge.

3. Set up the PCR machine with the touchdown PCR protocol described in Table 2. Begin the program and wait until the

Table 1
Master mix for microsatellite fragment analysis PCR

Reagent	Volume for each reaction (μL)
Invitrogen PCR SuperMix	22.5
dH$_2$O	1.0
CATT forward primer, 10 μM	0.5
CATT reverse primer, 10 μM	0.5
Genomic DNA, 2–25 ng/μL	0.5
Total volume	25

Table 2
Cycling conditions for microsatellite fragment analysis PCR

Step	Temperature (°C)	Time	Notes
1	95	5:00	
2	95	0:30	
3	54	0:30	
4	72	0:45	Steps 2–4 × 39
5	72	10:00	
6	10	Hold	

PCR block is hot before loading the samples. While PCR is running, prepare a 1% (w/v) agarose in 1× TBE gel with an appropriate number of wells to accommodate samples, controls, and size standard lanes.

4. After the PCR reaction is finished, separately mix a portion of the product with loading buffer, *do not mix loading buffer directly into the PCR product* or it will be unusable for downstream processing. Set up electrophoresis tank with 1% agarose gel in 1× TBE, load the gel with mixture and size standards, and run at 120 V for approximately 30 min. Make sure to protect remaining PCR product from light.

5. Stain gel with equal volume of 3× GelRed and gently shake for 30 min at room temperature, protected from light before visualization on a UV transilluminator. You should be able to see the anticipated product sizes = 340–352 bp. If you do not see bands within this range, or see extraneous bands, you may need to optimize the PCR conditions for your situation.

6. Prepare two dilutions of the PCR product, at 1:10 and 1:100 with H_2O, with a final volume of 10 μL in a clean MicroAmp 96-well Plate. Record your samples on a 96-well template document. Seal the plate with film and wrap with foil to protect from light.

7. Capillary Electrophoresis [29, 30]. This procedure may need to be performed by your genomics core facility staff (*see* **Note 4**). 1–2 μL of the diluted PCR product is mixed with 1 μL GeneScan 500 LIZ size standard and 12.5 μL of deionized formamide. The samples are mixed by pipetting up and down after addition of the PCR product.

8. The samples are denatured at 95 °C for 5 min and snap-cooled in an ice-water bath before being placed in the autosampler tray on the 3730xl Genetic Analyzer (Applied Biosystems) instrument for automatic injection into the capillary.

9. The samples are run on a 50 cm capillary array (Applied Bio-systems) and electrokinetically injected for 15 s at 15 kV, and then run at 15 kV for 27 min at a constant temperature of 63 °C.

3.2 Data Analysis of Microsatellite Fragment Length Assay

After obtaining your data files from the ABI Genetic Analyzer. You will need to use the GeneMapper software to analyze your results and determine the CATT repeat numbers. The exact procedure will vary widely between software versions and system configurations. We provide generalized instructions below and these should be adapted in consultation with your genomics core facility staff and/or software support staff.

1. Open and log into **GeneMapper** software. Choose **File** > **Add Samples to Folder** > Select the folder that your samples files are in > **Add to List** > **Add.**

2. If the following has not been done already, you will need to set up Size Standards, Kit, Bin Sets, Panel, Marker, and Bins according to the below instructions. Otherwise you may skip to **step 3**.

   For **Size Standard**, use GS500LIZ. If it is not present already in the software, use the **Size Standard Editor** to enter the sizes from the GeneScan-500 LIZ product documentation and select **Size Standard Dye**: Orange.

   In the **Panel Manager,** create a **New Kit** with the **Kit Name**: "CATT repeats", select the new created kit and add a **New Bin Set** with the **Name**: "CATT repeats". Create a **New Panel** with the **Panel Name**: "CATT". Select the newly created panel, and create a **New Marker** with the **Marker Name**: "CATT", **Dye Color**: "blue" (default), **Min Size**: 330, **Max Size**: 362

   Select the newly created marker, and **Add Multiple Bins**, entering the following information into the window, **Starting Size**: 340, **End Size**: 353, **Left Offset**: 0.4, **Right Offset**: 0.4, **Bin Spacing**: 4.0. Click **OK**. Rename the automatically numbered bins (1, 2, 3, 4) to correspond with the number of CATT repeats (5, 6, 7, 8). Do so by right clicking each individual bin number and select **Edit Bin**.

   Create a **New Analysis Method**, in the **General** tab add to **Name**: "CATT" and in the **Allele** tab select **Bin Set**: "CATT repeats".

3. In the first row of your list of samples, select the following option for each column:
   Analysis Method: "CATT".
   Panel: Click "MIF-CATT" folder and select "CATT" panel.
   Size Standard: GS500 LIZ.

   To fill down remaining rows with the above options, select and highlight the whole column by clicking the column name.

Go to **Edit** > **Fill Down**. Select all the rows and click **Analyze** (green arrow button).

4. Go to the Genotypes tab. You will see your results. CATT-5 (340 bp); CATT-6 (344 bp); CATT-7 (348 bp); CATT-8 (352 bp). If the peak height is >1000, the result is probably good and the allele (i.e., number of repeats) will be automatically called.

5. It is important for the user to visually check the peaks in the **Plot View** for all samples to verify correct binning into each allele. You may also get lower peaks for which you will have to manually call the allele present or exclude the sample altogether if it appears to be a false peak (*see* **Note** 7). Lastly, check for concordant results between the 1:10 and 1:100 dilutions of your PCR product.

3.3 Genotyping the MIF – 173 G/C SNP by TaqMan Allelic Discrimination Assay

1. Thaw, mix well, and centrifuge the reagents and samples. Make sure to keep probe on ice and not exposed to light (it is optional, but recommended, to dilute DNA 1:1 with ddH$_2$O. 3.2 μL of each is sufficient for two reps). Make the dilutions in 96-well plates, as this will permit use of a multichannel pipettor for ease of transfer to 384-well plates.

2. Prepare a master mix according to Table 3, without genomic DNA, on ice. Make 15% more mix than is necessary to allow for pipetting error. Mix well and spin down in centrifuge. Load 2.63 μL of master mix into each well on the 384-well clear optical reaction plate.

3. Add genomic DNA, changing tips between each well. Include at least one NTC (no template control) well for each row. The NTC wells are ones in which all reagents are added but no DNA sample is included because ddH$_2$O is added instead. Record your samples on a 384-well template document.

4. Cover with optical adhesive covers. Note: Use spatula to make sure cover is sealed. Ensure the cover is not damaged or dirty

Table 3
Master mix for TaqMan allelic discrimination assay

Reagent	Volume for each reaction (μL)
TaqMan Genotyping Master Mix, 2×	2.5
Assays-On-Demand SNP Genotyping Assay Mix, 40× (*see* Note 5)	0.13
Genomic DNA, diluted in ddH$_2$O	2.37
Total volume	5

because this will interfere with the read. Mix well and spin down in centrifuge. Wrap the plate in aluminum foil to protect from light, keep on ice until ready to read.

3.4 Pre-run Read for TaqMan Allelic Discrimination Assay

Do not skip the pre-run read because it is necessary for optimal results by allowing the software to subtract the pre-run background from the plate. We use the ABI 7900HT Real-Time Fast PCR machine with SDS 2.4 software. The exact procedure will vary widely between software versions and system configurations. We provide generalized instructions below and these should be adapted in consultation with your genomics core facility staff and/or software support staff.

1. Create a new **Allelic Discrimination** plate document.

2. If this is your first time running you will need to create **New Marker** in the **Marker Manager** and label it "MIF-GC," otherwise you may skip to Subheading 3.4, **step 3**. You will also need to create **New Detectors** for MIF-C (reporter = VIC) and MIF-G (reporter = FAM) in the **Detector Manager**. Add both detectors to the marker you created, set MIF-C as **Allele X** with **Base = C**, and MIF-G as **Allele Y** with **Base = G**.

3. Click **Add Marker**, select the marker and then click **Copy to the Plate**. Select the wells containing samples and check the **Use** box on the marker, make sure "Unknown" is selected as the task. Select the wells containing NTC, check Use box on the marker and select "NTC" as the task.

4. **Save** and give your new file a name (this will also be the post-read document). Select the **Instrument** tab and **Plate Read** sub-tab in the right screen, click the **Connect** button. When connected, click the **Open/Close** button in the same tab. Put your plate into the machine. Make sure the A1 on the plate is oriented to the A1 on the machine. For the plate read, in the **Instrument** tab, click **Pre Read**. Machine will close automatically. If successful, a Date Collection Stamp will appear next to "Pre:". **Save** the file.

3.5 Running the TaqMan Allelic Discrimination Assay PCR

1. Create a new **Standard Curve (AQ)** document.

2. Click **Add Detectors**, add both "MIF-G" and "MIF-C" (return to Subheading 3.4, **Step 2** if these have not been set up already) and **copy to the plate**. Select all wells being used, both samples and NTC, and check the **Use** box for both "MIF-G" and "MIF-C" detectors.

3. Select the **Instrument** tab in the right screen. In the Thermal Profile sub-tab, next to mode, select the **9600 Emulation** radio button. Set the **Sample Volume** to 5 μL and the number of **Repeats** to 50. Enter the PCR parameters described in Table 4.

Table 4
Cycling conditions for TaqMan allelic discrimination assay

Step	Temperature (°C)	Time	Notes
1	95	10:00	
2	92	0:15	
3	60	1:00	Steps 2–3 × 50

4. Save this file and give it a separate name from your Allelic Discrimination document.

5. Go to **Real Time** sub-tab under the Instrument tab. If not connected, click **Connect** button. Click **Start Run** button. Wait for a time to appear at "Run Time Remaining:" under instrument status. You can leave for the duration of the real-time PCR.

3.6 Post-run Read and Analysis for TaqMan Allelic Discrimination Assay

1. Reopen your Allelic Discrimination document file (where you collected and saved your pre-read data for this plate).

2. Select **Instrument** tab and the **Plate Read** sub-tab. If machine is disconnected, click **Connect**. For the post-run plate read, click **Post Read**. A date stamp should appear.

3. Save your file. Click **Open/Close** button and remove your plate. Click **Open/Close** again to close. Click **Disconnect**.

4. Click the green **Analyze** button. Under the **Results** tab, you can look at your samples and the allele calls that were made for the MIF–173 SNP locus: remember **MIF-G = FAM = G** allele and **MIF-C = VIC = C** allele. If your calls are undetermined you may have to enable 2-cluster analysis (*see* **Note 8**). Manual calling is also possible, please consult the full guide (*see* **Note 9**). Alternatively, samples which result in no genotype calls using this protocol can be used for the MIF SNP PCR-RFLP protocol as a secondary assay. In our experience, we have been able to obtain further SNP genotyping data from a subset of samples that were not successfully genotyped by the TaqMan assay. However, we do not recommend using PCR-RFLP as the primary protocol in the genotyping workflow because of its difficulty for processing large numbers of samples and labor intensiveness.

5. If you are satisfied with the analysis and the allele calling, save your file to keep your analysis settings and the calling data. You can transfer the files to another computer for analysis if you have the SDS software (*see* **Note 10**). Otherwise, you can export the results as a text file from the file menu, or export images of the grid or plot.

Table 5
Master mix for PCR-RFLP amplification

Reagent	Volume for each reaction (μL)
ddH₂0	14.88
5× KAPA HotStart Buffer	5.0
MgCl₂ (from kit)	1.5
Forward primer (10 μM)	0.5
Reverse primer (10 μM)	0.5
dNTPs (from kit)	0.5
KAPA HotStart *Taq* polymerase enzyme	0.12
Template DNA	2
Total volume	25

3.7 Genotyping the MIF −173 G/C SNP by PCR-RFLP Analysis

1. Thaw, mix well, and centrifuge all reagents and samples before use. Prepare a PCR master mix with the KAPA *Taq* HotStart PCR Kit, on ice, per Table 5, without the template DNA. Add the *Taq* polymerase to the master mix last. Mix well and spin down in centrifuge. Prepare 15–20% more master mix than needed (for example, when preparing master mix for 10 samples and controls, make enough master mix for 12 reactions).

2. Aliquot 23 μL of the master mix into each PCR tube. Add 2 μL of genomic DNA into each reaction well, changing tips between every well. For the negative control well, add 2 μL of ddH₂O instead. The final reaction volume will be 25 μL. Check to make sure caps are on securely, ensure your samples are thoroughly mixed and spin down in centrifuge.

3. Set up the PCR machine with the PCR cycling conditions described in Table 6. This touchdown PCR protocol allows both greater specificity and robustness of amplification versus standard PCR protocols. Begin the program and wait until the PCR block is hot before loading the samples. While PCR is running, prepare a 1% (w/v) agarose in 1× TBE gel with an appropriate number of wells to accommodate samples, controls, and size standard lanes.

4. Verify amplification of the correct product by running the PCR product on a 1% in 1× TBE agarose gel. Do not add loading buffer directly to PCR product, instead take 5 μL of the PCR product and mix with 2.5 μL 6× loading buffer in new tubes. Set up electrophoresis tank, load the gel with 5 μL of this mixture, load the size standards, and run at 120 V for approximately 30 min.

Table 6
Cycling conditions for PCR-RFLP amplification

Step	Temperature (°C)	Time	Notes
1	94	5:00	
2	94	0:30	
3	57	0:40	−0.5 °C per cycle
4	72	0:40	Steps 2–4 × 12
5	94	0:30	
6	51	0:40	
7	72	0:40	Steps 5–7 × 25
8	72	2:00	
9	10	Hold	

Table 7
Master mix for PCR-RFLP enzyme digest

Reagent	Volume (μL)
10× CutSmart Buffer (*see* Note 12)	0.5
ddH$_2$O	3
NEB *AluI* enzyme	0.5
Amplified PCR product	6
Total volume	**10**

5. After running the gel, stain with equal volume of 3× GelRed (prepared in water, following post stain protocol) for approximately 30 min with gentle shaking. Image the gel, a clear band should be present around 366 base pairs (the length of the target).

6. After verifying successful amplification, prepare a master mix for the *AluI* enzyme digest according to Table 7, without the PCR product (*see* **Note 11**). Keep *AluI* restriction enzyme in an enzyme block while working (it does not need to be thawed). Prepare 15–20% more master mix than needed.

7. Aliquot 4 μL of master mix, then add 6 μL PCR product in new PCR tubes. Mix thoroughly by pipetting, then briefly centrifuge.

8. Load the reaction mix into a PCR machine and run the incubation program described in Table 8. While the digest samples are incubating, prepare a 3% (w/v) low-melting point agarose

Table 8
Incubation conditions for PCR-RFLP enzyme digest

Step	Temperature (°C)	Time
1 (*see* Note 13)	37	6:00:00
2	80	20:00
3	4	Hold

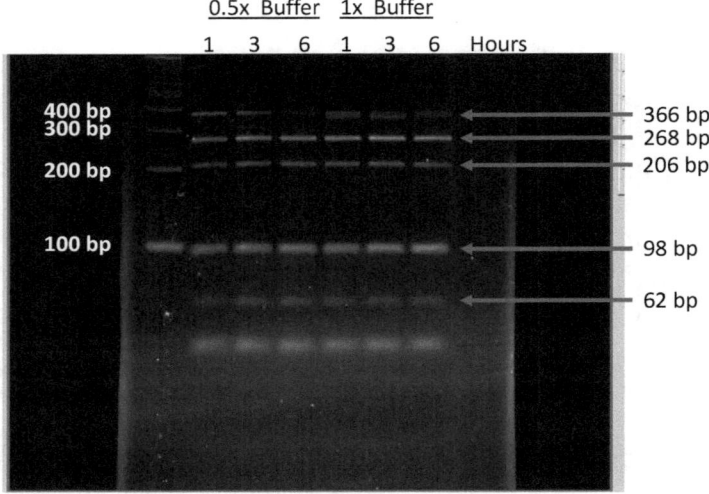

Fig. 3 Optimization of AluI restriction digest conditions. The first lane is a 100 bp DNA marker. Subsequent lanes contain digested PCR product from a GC heterozygote. The undigested portion of PCR product is faintly visible at 366 bp. The digested product from the GC heterozygote consists of four fragments: 268, 206, 98, and 62 bp. Longer incubation results in more complete digestion and using the enzyme buffer at 0.5× concentration improved the result (*see* **Notes 12** and **13**)

in 1× TBE gel with an appropriate number of wells to accommodate samples, controls, and size standard lanes.

9. After digest completion, add 2.5 μL of 6× Loading Buffer to each well. Set up electrophoresis tank with 3% low-melting point agarose in 1× TBE gel, load 10 μL of the mixture, and load the size standards. Start run at constant 80 V for 5 min and then constant 120 V, running until dye front is ¾ of the way down the gel.

10. Once the gel is finished running, stain the gel with 3× GelRed in water for approximately 30 min with gentle shaking before imaging the gel.

11. Analysis of PCR-RFLP results: As previously stated, the pre-digest PCR product should present a band at 366 base pairs. Following digest of the PCR product with *AluI* you should be able to detect the following bands which will indicate the SNP present at the MIF—173 locus (Fig. 3):

    GG homozygous—two fragments at 268 bp, 98 bp

    CC homozygous—three fragments at 206 bp, 98 bp, 62 bp

    GC heterozygous—four fragments at 268 bp, 206 bp, 98 bp, 62 bp

4   Notes

1. For extraction from blood or plasma, often one must use a suitable anticoagulation reagent; however heparin can interfere with the PCR reaction by inhibiting DNA polymerase, so sodium citrate or EDTA is recommended [24, 25].

2. If the samples only have trace of gDNA (less than 1 ng/μL) you may wish to amplify it using a whole genome amplification (WGA) kit. This technology is based on multiple displacement amplification (MDA) using DNA polymerase from *Bacillus subtilis* phage Phi29 and elongates templates up to 70,000 bp independently of sequence composition. We recommend the REPLI-g mini kit (QIAGEN) [26, 27].

3. Other brands of enzymes and plasticware may work for this assay; however we have not tested other brands. The product used needs to be compatible with the chemistry on the genetic analyzer being used and we have listed the exact brand that works for our setup. Further optimization will be required if another brand is selected.

4. We use a 6-FAM (6-Carboxyfluorescein) fluorescent dye attached to the 5′ end of the reverse primer. This fluorescent dye is most commonly used for attachment to oligonucleotides and should be compatible with most equipment. Nonetheless, the selection of an appropriate fluorescent label should be done in consultation with the staff who operate the genetic analyzer at your institution's sequencing facility because it will depend on the chemistry of the machine being used. We have listed the specific reagents needed for the machine mentioned; however at our institution, the machine-specific reagents (capillary, size standards, formamide) are supplied by the facility and do not typically need to be purchased by the user. Furthermore, the final preparation of the labeled PCR products and loading onto the capillary electrophoresis machine are generally performed by the facility staff rather than the user.

5. The genotyping master mix is optimized for use with the listed genotyping assay mix, which is a pre-mixed assay reagent that includes both MIF primers and the fluorescently labeled probes (FAM & VIC). The genotyping assay mix, which is essential to performing the TaqMan assay, has a proprietary design only available from Applied Biosystems and we therefore recommend using the product number and custom design number mentioned in the materials section. We are not currently aware of any products from other companies that can be substituted for this kit. Plasticware may be substituted if it is compatible and we have listed the brand of plastics that work with the machine mentioned.

6. We use this specific kit and brand of enzyme from KAPA because we found it has the best success with difficult to amplify samples after trialing several other kits. This is important for us since we use this protocol as a secondary assay for samples that don't work with the primary TaqMan assay. We also found that the pre-mix version of the same PCR kit did not perform as well.

7. There are many ways to look at the peaks and the quality of the data. Please consult the user guide, "DNA Fragment Analysis by Capillary Electrophoresis" (Applied Biosystems, publication number 4474504).

8. 2-cluster calling is enabled by selecting the **Analysis Settings** from the **Analysis** drop-down menu. Within the **Analysis Settings** dialog box, select the **Marker** tab, select both the **Auto caller enable** and the **2-cluster calling enabled** boxes. In the **Quality Value** field, you may change the percentage value to apply as quality interval (the greater the value, the more stringent the calls), otherwise leave it alone.

   If still no calls are made—check to see if any of the wells have errors. If there are any error-ridden wells, you will have to close out of the current allelic discrimination document. Reopen the allelic discrimination document, and in the Setup tab, select all the wells with errors and exclude them from the analysis. Then continue through the rest of the analysis steps.

9. A full guide is available for the Applied Biosystems 7900HT Fast Real-Time PCR System called the "Allelic Discrimination Getting Started Guide" (Applied Biosystems #4364015).

10. The SDS software can be downloaded for use on another computer for data analysis. Please contact Thermo-Fisher or go to their website for the download registration form. You will need to enter your serial number from your ABI 7900HT real-time machine to obtain a free software license.

11. Open the amplified PCR product in a different location than where you prepare PCRs to avoid cross-contamination. This is

especially important while setting up the enzyme digest. Make sure that you have prepared the master mix and aliquoted it before you start opening the PCR product tubes and adding it to the digest wells.

12. Enzyme buffer is added at $0.5\times$ because there is residual buffer from the PCR product which will be already present in the mixture and adding $1\times$ buffer will lead to suboptimal digestion (Fig. 3).

13. There is usually some undigested PCR product left over after the restriction digest. This undigested product will persist as a band that can be detected at approximately 366 bp. Longer digest times will decrease the amount of undigested product, there is usually no undigested product after an overnight digest.

Acknowledgements

This work was supported by National Institutes of Health grants AR049610 and HL130669 to R.B. We are grateful to the staff of the W. M. Keck Foundation DNA Sequencing Facility at Yale School of Medicine for their assistance in developing the microsatellite fragment analysis and TaqMan assay protocols. We are also grateful to Bruce Liberi for his assistance in developing and optimizing the PCR-RFLP protocol.

References

1. Calandra T, Roger T (2003) Macrophage migration inhibitory factor: a regulator of innate immunity. Nat Rev Immunol 3:791–800. https://doi.org/10.1038/nri1200

2. Morand EF, Leech M, Bernhagen J (2006) MIF: a new cytokine link between rheumatoid arthritis and atherosclerosis. Nat Rev Drug Discov 5:399–410

3. Yende S, Angus DC, Kong L et al (2009) The influence of macrophage migration inhibitory factor (MIF) polymorphisms on outcome from community-acquired pneumonia. FASEB J 23:2403–2411

4. Calandra T, Bernhagen J, Metz CN et al (1995) MIF as a glucocorticoid-induced modulator of cytokine production. Nature 377:68–71. https://doi.org/10.1038/377068a0

5. Mitchell RA, Liao H, Chesney J et al (2002) Macrophage migration inhibitory factor (MIF) sustains macrophage proinflammatory function by inhibiting p53: regulatory role in the innate immune response. Proc Natl Acad Sci U S A 99:345–350

6. Roger T, David J, Glauser MP, Calandra T (2001) MIF regulates innate immune responses through modulation of Toll-like receptor 4. Nature 414:920–924

7. Baugh JA, Chitnis S, Donnelly SC et al (2002) A functional promoter polymorphism in the macrophage migration inhibitory factor (MIF) gene associated with disease severity in rheumatoid arthritis. Genes Immun 3:170–176

8. Zhong X, Leng L, Beitin A et al (2005) Simultaneous detection of microsatellite repeats and SNPs in the macrophage migration inhibitory factor (MIF) gene by thin-film biosensor chip and application to rural field studies. Nucleic Acids Res 33:2121

9. Renner P, Roger T, Bochud PY et al (2011) A functional microsatellite of the macrophage migration inhibitory factor gene associated with meningococcal disease. FASEB J 26:907–916

10. De Benedetti F, Meazza C, Vivarelli M et al (2003) Functional and prognostic relevance of the −173 polymorphism of the macrophage migration inhibitory factor gene in systemic-onset juvenile idiopathic arthritis. Arthritis Rheum 48:1398–1407

11. Radstake TR, Sweep FC, Welsing P et al (2005) Correlation of rheumatoid arthritis severity with the genetic functional variants and circulating levels of macrophage migration inhibitory factor. Arthritis Rheum 52:3020–3029

12. Wu SP, Leng L, Feng Z et al (2006) Macrophage migration inhibitory factor promoter polymorphisms and the clinical expression of scleroderma. Arthritis Rheum 54:3661–3669

13. Das R, Loughran K, Murchison C et al (2016) Association between high expression macrophage migration inhibitory factor (MIF) alleles and West Nile virus encephalitis. Cytokine 78:51–54

14. Sreih AG, Ezzeddine R, Leng L et al (2011) Dual effect of MIF gene on the development and the severity of human systemic lupus erythematosus. Arthritis Rheum 63:3942–3951

15. Meyer-Siegler KL, Vera PL, Iczkowski KA et al (2007) Macrophage migration inhibitory factor (MIF) gene polymorphisms are associated with increased prostate cancer incidence. Genes Immun 8:646–652

16. Grigorenko EL, Han SS, Yrigollen CM et al (2008) Macrophage migration inhibitory factor and autism spectrum disorders. Pediatrics 122:e438–e445

17. Barton A, Lamb R, Symmons D et al (2003) Macrophage migration inhibitory factor (MIF) gene polymorphism is associated with susceptibility to but not severity of inflammatory polyarthritis. Genes Immun 4:487–491

18. Berdeli A, Mir S, Ozkayin N et al (2005) Association of macrophage migration inhibitory factor −173C allele polymorphism with steroid resistance in children with nephrotic syndrome. Pediatr Nephrol 20:1566–1571

19. Shimizu T, Hizawa N, Honda A et al (2005) Promoter region polymorphism of macrophage migrationinhibitory factor is strong risk factor for young onset of extensive alopecia areata. Genes Immun 6:285–289

20. Zhang H, Ma L, Dong LQ et al (2013) Association of the macrophage migration inhibitory factor gene—173G/C polymorphism with inflammatory bowel disease: a meta-analysis of 4296, subjects. Gene 526:228–231

21. Guttmacher A, Collins FS (2002) Genomic medicine—a primer. N Engl J Med 347:1512–1520

22. Wang FF, Huang XF, Shen N et al (2013) A genetic role for macrophage migration inhibitory factor (MIF) in adult-onset Still's disease. Arthritis Res Ther 15:R65. https://doi.org/10.1186/ar4239

23. Ramireddy L, Lin CY, Liu WY et al (2014) Association study between macrophage migration inhibitory factor-173 polymorphism and acute myeloid leukemia in Taiwan. Cell Biochem Biophys 70:1159–1165. https://doi.org/10.1007/s12013-014-0036-z

24. Yokota M, Tatsumi N, Nathalang O et al (1999) Effects of heparin on polymerase chain reaction for blood white cells. J Clin Lab Anal 13:133–140

25. Holodniy M, Kim S, Katzenstein D et al (1991) Inhibition of human immunodeficiency virus gene amplification by heparin. J Clin Microbiol 29:676–679

26. Lipschutz R, Bick J, Nguyen V et al (2018) Macrophage migration inhibitory factor (MIF) gene is associated with adolescents' cortisol reactivity and anxiety. Psychoneuroendocrinology 95:170–178

27. Bäumer C, Fisch E, Wedler H et al (2018) Exploring DNA quality of single cells for genome analysis with simultaneous whole-genome amplification. Sci Rep 8:7476

28. Blanco L, Bernad A, Lázaro JM et al (1989) Highly efficient DNA synthesis by the phage phi 29 DNA polymerase. Symmetrical mode of DNA replication. J BiolChem 264:8935–8940

29. Thermo Fisher Scientific Inc. (2014) DNA fragment analysis by capillary electrophoresis. Applied Biosystems Publication Number 4474504, Revision B

30. Lazaruk K, Walsh PS, Oaks F et al (1998) Genotyping of forensic short tandem repeat (STR) systems based on sizing precision in a capillary electrophoresis instrument. Electrophoresis 19:86–93

Chapter 8

Staining MIF in Cells for Confocal Microscopy

James Harris

Abstract

Confocal microscopy is a powerful technique for immunofluorescence imaging of cells and tissues. The technique allows for detailed analysis of intracellular localization of molecules, as well as three-dimensional representation and analysis of samples, and can be used as a gateway to more advanced techniques, including FLIM-FRET and super-resolution microscopy. Relatively few studies have used confocal microscopy to study intracellular localization of macrophage migration inhibitory factor (MIF) in detail. This chapter outlines basic protocols and tips for staining MIF in fixed cells for confocal analysis.

   Key words Antibody staining, Confocal, Fluorophores, Immunofluorescence

1   Introduction

Confocal microscopy is an invaluable tool for a broad range of biological investigations. Compared to conventional wide-field immunofluorescence microscopy, confocal microscopy offers many benefits. A confocal microscope selectively captures light from a thin (<1 μm) optical section at the plane of focus. As a result, confocal microscopy allows the user to control the depth of field, reducing out-of-focus light and providing more detailed information on specific localization (and potential co-localization) of target molecules. In addition, serial optical sections can be obtained, effectively revealing the three-dimensional structure of a specimen [1]. Confocal/immunofluorescence imaging can be performed on tissue sections, whole organisms, in vivo/intravital specimens (*see* Chapter 3) and cell monolayers and matrices. Confocal microscopy can be performed on both fixed and live specimens, thus adding a fourth dimension. Moreover, in many cases, samples prepared for confocal microscopy can also be used for other techniques, including fluorescence lifetime imaging-fluorescence/Förster resonance energy transfer (FLIM-FRET) and super-resolution

James Harris and Eric F. Morand (eds.), *Macrophage Migration Inhibitory Factor: Methods and Protocols*,
Methods in Molecular Biology, vol. 2080, https://doi.org/10.1007/978-1-4939-9936-1_8,
© Springer Science+Business Media, LLC, part of Springer Nature 2020

microscopy, although in some cases specific fluorophores are required for optimum analysis (*see* Chapter 9).

Originally characterized as a cytokine, macrophage migration inhibitory factor (MIF) is a pleiotropic immunomodulatory molecule implicated in numerous disease states [2]. Surprisingly few studies have looked at MIF in cells by confocal microscopy, so relatively little detail is known about its intracellular localization. Thus, there is a clear opportunity to gain further insight into MIF biology through the use of confocal microscopy. A recent report used confocal microscopy to demonstrate that MIF co-localizes with NLRP3 and the NLRP3 inflammasome [3]. In addition, the same study used related FLIM-FRET analysis (*see* Chapter 9) to show potential interaction between MIF and NLRP3, as well as NLRP3 and the intermediate filament protein vimentin. Another study has demonstrated co-localization of MIF and the Golgi-associated protein p115 [4]. In both cases, the distribution of MIF in macrophages was shown to be widespread throughout the cytosol and in as yet unidentified structures/vesicles (Fig. 1).

This chapter outlines the basic methods for staining cells for confocal microscopy-based analysis of MIF. Included are hints and tips for optimization of protocols. Careful specimen preparation and selection of the best reagents are critical for successful confocal microscopy. While this technique makes images sharper and more detailed, it cannot mitigate poor sample preparation or inferior reagents. It is highly recommended that the user spend time finding the best antibody for their cells and ensuring its specificity. I have found that some MIF antibodies will work well for Western blot, but not confocal, while others will work in human, but not mouse cells (and vice versa). Importantly, I have also found one antibody that stains something other than MIF in a highly specific and potentially misleading way (Fig. 1b). Thus, testing antibodies in $Mif^{-/-}$ cells is highly recommended (note that while some companies now offer knockout-validated antibodies, it is still important to test them fully). The techniques outlined here are for macrophages, but can be applied to any adherent cells, with appropriate optimization.

2 Materials

2.1 Preparing Coverslips for Macrophage Adherence

1. Cover glass II (or equivalent) or #1.5 (0.17 mm) glass coverslips (*see* **Note 1**).

2. Concentrated nitric acid (70%).

3. Methanol.

4. MilliQ H_2O.

5. Glass container, with lid (autoclavable).

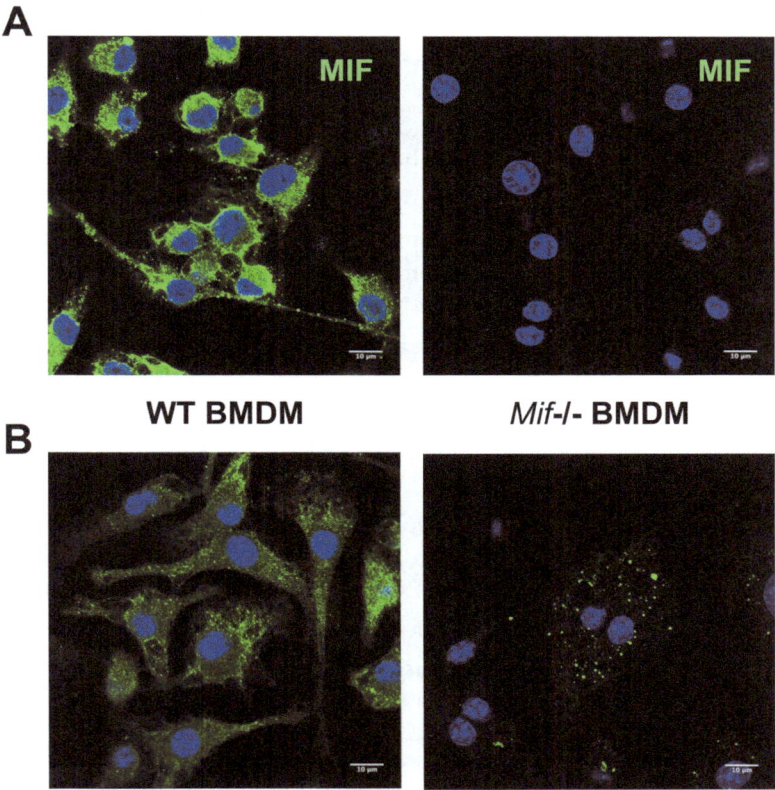

Fig. 1 Confocal analysis of MIF staining in mouse bone marrow-derived macrophages (BMM). (**a**) Wild-type (WT) and $Mif^{-/-}$ BMM were treated with LPS (100 ng/mL) overnight, fixed in 2% paraformaldehyde, and stained with a rabbit polyclonal antibody against MIF. MIF can be seen in punctate structures, as well as more diffusely localized in the cytosol. This antibody did not stain $Mif^{-/-}$ cells. (**b**) WT and $Mif^{-/-}$ BMM were treated with LPS (100 ng/mL) overnight, fixed in 2% paraformaldehyde, and stained with a different rabbit polyclonal antibody against MIF. While the cytosolic staining in the WT cells is comparable to (**a**), significant punctate staining can be seen in the $Mif^{-/-}$ cells, suggesting nonspecific staining

2.2 Culturing Macrophages on Coverslips

1. Cells of choice.

2. Complete medium: RPMI-1640, supplemented with 10% FCS, 2 mM L-glutamine, 50 U/mL penicillin, and 50 μg/mL streptomycin (*see* **Note 2**).

3. Prepared coverslips (*see* Subheading 3.1).

4. 12-well tissue culture plates.

2.3 Fixing Cells

1. Phosphate buffered saline (PBS); 137 mM NaCl, 10 mM Na_2HPO_4, 1.8 mM KH_2PO_4, 2.7 mM KCl in milliQ water,

adjust to pH 7.4 with HCl. Sterilize by autoclaving or filtration.

2. 2% (w:v) paraformaldehyde in PBS.

2.4 Staining Cells for Immunofluorescence

1. PBS.

2. Triton X-100 0.1% (v:v) in PBS.

3. Immunofluorescence (IF) blocking solution; 1% bovine serum albumin (BSA) in PBS with 5% serum from the same species as the secondary antibodies (e.g., goat, donkey). Filter with a 0.2 μm syringe tip filter.

4. Primary antibody/antibodies, diluted 1/100–1/2000 in IF blocking solution (*see* **Note 3**).

5. Fluorescently labeled secondary antibody/antibodies, diluted 1/1000–1/5000 in IF blocking solution (*see* **Note 4**).

6. Nuclear stain; e.g., DAPI (2-(4-Amidinophenyl)-6-indolecarbamidine dihydrochloride), final concentration of 1–5 μg/mL in PBS (from a stock of 5 mg/mL in PBS) (*see* **Note 5**).

7. Fluorescent mounting medium (*see* **Note 6**).

8. Glass slides.

3 Methods

3.1 Preparing Coverslips for Macrophage Adherence

1. Soak coverslips in concentrated nitric acid (in fume hood) in a glass container with lid for 1 h.

2. Rinse 10 times with 1 L Milli-Q water.

3. Incubate in methanol overnight.

4. Remove methanol and leave to dry in fume hood or drying cabinet until completely dry (*see* **Note 7**).

5. Sterilize by autoclave.

3.2 Culturing Macrophages on Coverslips

1. Using sterile forceps, transfer coverslips to the wells of a 12-well tissue culture plate.

2. To each well, add cells at 5×10^5 cells/mL, 1 mL per well. After pipetting, gently push down on each coverslip with the tip of the pipette, to push out air bubbles from under the coverslip (*see* **Note 8**).

3. Allow the cells to adhere for 2–24 h before treatments.

3.3 Fixing Cells

1. After treatments, remove medium from cells and wash once with cold PBS.

2. Remove PBS and replace with 2% paraformaldehyde. Incubate at room temperature for 30 min.

3. Wash three times with PBS, finally leaving the cells in 1 mL PBS. At this point, cells can be kept at 4 °C for up to a month before staining (*see* **Note 9**).

3.4 Staining Cells for Immunofluorescence

1. Permeabilize the cells with 0.1% Triton X100 (500 µL per well) for 10 min at room temperature (*see* **Note 10**).

2. Wash the coverslips three times with PBS.

3. Block the cells with IF blocking solution (500 µL per well) for 1 h at room temperature.

4. Wash the coverslips three times with PBS (1 mL per well).

5. Incubate cells with primary antibody/antibodies (*see* **Note 11**) at room temperature for 1–2 h. To minimize antibody wastage, we use 50 µL of antibody solution dropped on parafilm stretched over the base of a tissue culture plate and the coverslip is then put on this (cell side down) (Fig. 2) (*see* **Note 12**).

6. Transfer coverslips back to the 12-well plate (cell side up). Wash the coverslips three times with PBS (1 mL per well).

7. Add secondary, fluorescently labeled, antibodies (300–500 µL per well) and incubate at room temperature for 1 h (*see* **Note 13**).

8. If required, stain nuclei with appropriate stain (e.g., DAPI) for 5–10 min at room temperature.

9. Wash the coverslips three times with PBS.

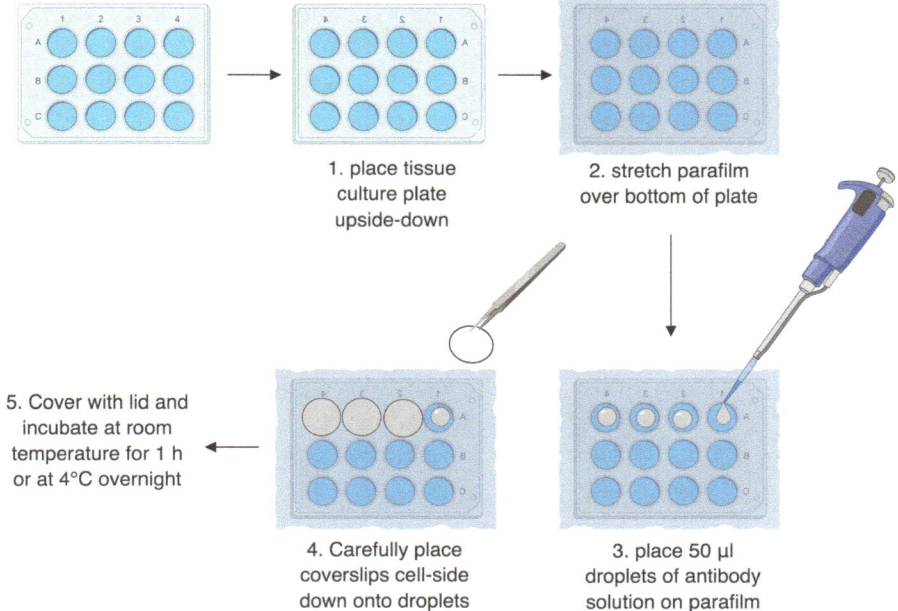

1. place tissue culture plate upside-down

2. stretch parafilm over bottom of plate

3. place 50 µl droplets of antibody solution on parafilm

4. Carefully place coverslips cell-side down onto droplets

5. Cover with lid and incubate at room temperature for 1 h or at 4°C overnight

Fig. 2 Protocol for staining coverslips on parafilm

10. Mount the coverslips onto slides with fluorescent mounting medium.

11. Image as required.

4   Notes

1. #1.5 coverslips (0.17 mm thickness) are required to obtain the best (brightest, crispest, high resolution) images by confocal/immunofluorescence microscopy, so should always be used. Different sizes can be used, depending on the culture plates/dishes used. For example, round 18 mm diameter coverslips fit well in 12-well plates.

2. For most macrophage cultures, we use RPMI-1640 as described. However, other media can be used—choose the best for your cell type/application.

3. Optimization of antibody concentration is recommended for each antibody and cell type.

4. Optimization of antibody concentration is recommended.

5. Many different nuclear stains are available for confocal and immunofluorescence microscopy. The best for any application will depend on multiple factors, including other dyes used on samples and lasers/filters available. Excitation/Emission for DAPI is 358/461 nm, so a 405 nm diode laser is optimal. Other dyes (with excitation\emission) include TO-PRO-3™ (642/641 nm), SYTO™ dyes (various spectra), SYTOX™ (various spectra, for fixed/dead cells), and Hoechst 33342 (350/461 nm).

6. You can tell when the coverslips are completely dry by shaking the container they are in—when completely dry, individual coverslips will move freely about but if still wet, they will stick together.

7. We routinely use 2% paraformaldehyde in PBS to fix cells. While this works for most antibodies and applications, we advise optimizing for each new antibody/application. Alternatives are ice-cold methanol for 5 min or 3.7–4% paraformaldehyde. Some antibodies work better under specific fixation and methanol can be better for maintaining morphology for some applications, such as super-resolution microscopy, but may not be appropriate for membrane proteins.

8. This will stop cells growing on the underside of the coverslip and will make the coverslips easier to remove later.

9. Ideally, cells should be stained and imaged as soon after fixation as possible. However, they can be left after fixing (which is preferable to leaving after staining) for at least a month

(we have left them longer). Sample quality will deteriorate over time, though.

10. Saponin (0.1% in PBS) can be used as an alternative for permeabilization. Some protocols permeabilize for longer (30–60 min) and others include Triton X-100 or saponin in all subsequent buffers. This can be optimized for each antibody/application.

11. Cells can be labeled with more than one primary antibody at a time, as long as they are from different species (note that most fluorescent secondary antibodies for confocal are species-specific, rather than isotype-specific).

12. An alternative is to use chamber slides, which allow for multiple samples to be cultured and stained on a single slide.

13. At this stage, other reagents, such as fluorescently tagged phalloidin for detecting F-actin, or membrane-staining dyes, can be added as well.

References

1. Hanrahan O, Harris J, Egan C (2011) Advanced microscopy: laser scanning confocal microscopy. Methods Mol Biol 784:169–180

2. Harris J et al (2019) Rediscovering MIF: new tricks for an old cytokine. Trends Immunol. https://doi.org/10.1016/j.it.2019.03.002

3. Lang T et al (2018) Macrophage migration inhibitory factor is required for NLRP3 inflammasome activation. Nat Commun 9(1):2223

4. Merk M et al (2009) The Golgi-associated protein p115 mediates the secretion of macrophage migration inhibitory factor. J Immunol 182 (11):6896–6906

Chapter 9

Microscopy Methods for Imaging MIF and Its Interaction Partners

Kirstin D. Elgass, Sarah J. Creed, and Ina Rudloff

Abstract

Fluorescence microscopy has become a powerful tool to investigate proteins in their natural environment. Well-established techniques like widefield and confocal fluorescence microscopy have commonly been used for decades to visualize biomolecules in single cells and tissue sections. Live cell microscopy allows for the investigation of biomolecular trafficking, and other specialized techniques, such as proximity ligation assays (PLA) and fluorescence lifetime imaging microscopy (FLIM), can be used to study interactions between biomolecules of interest. Finally, with the most recent rise of optical super-resolution microscopy, we can investigate target biomolecules in situ with unprecedented detail on the nanometer scale. Here, we discuss various optical microscopy techniques that have successfully been used to image MIF. We highlight applications, advantages, and limitations of each technique. The techniques described here can easily be adapted to investigate other target proteins, their localization, interaction partners, and mechanisms of action.

Key words Fluorescence, Confocal microscopy, FLIM, PLA, Super-resolution microscopy

1 Introduction

To understand the function of MIF, it is crucial to visualize MIF proteins and their interaction partners in their natural environment. Advanced optical fluorescence microscopy techniques are appropriate tools for examining protein interactions as specific proteins of interest can be fluorescently tagged and visualized directly in situ without the need for cell lysis and protein extraction. Excellent reviews and books on optical microscopy for biologists are available to the interested reader for more details [1–4]. Here we will use murine bone marrow-derived macrophages (BMMs) as a representative cell system to describe several of the optical fluorescence microscopy methods that can be used to examine MIF and its interaction partners. BMMs are relatively easy to obtain and culture and as macrophages express MIF endogenously, BMMs are an ideal model system [5].

James Harris and Eric F. Morand (eds.), *Macrophage Migration Inhibitory Factor: Methods and Protocols*, Methods in Molecular Biology, vol. 2080, https://doi.org/10.1007/978-1-4939-9936-1_9,
© Springer Science+Business Media, LLC, part of Springer Nature 2020

Optical fluorescence microscopy can be performed in fixed or live cells. When using fixed cells, target proteins can be detected with suitable antibodies, either directly by using fluorescently tagged primary antibodies or indirectly by using unconjugated primary antibodies and corresponding fluorescently labeled secondary antibodies. However, while this approach detects endogenous proteins and thus is free from potential overexpression artifacts, it does limit the information that can be obtained for protein dynamics and trafficking to end point analysis only.

For visualization of protein dynamics, DNA expression vectors encoding for the protein of interest linked to small fluorescent proteins, such as green fluorescent protein (GFP) [6], can be designed. Transfection of target cells with these vectors causes overexpression of the fluorescently labeled protein of interest, allowing localization and dynamics of the fluorescently tagged proteins to be imaged in the living cell in real time. Live cell imaging can also be carried out using cell-permeable dyes, such as cell masks or DAPI. Genetically encoded fluorescent proteins and cell-permeable dyes can be used in combination, provided different fluorophores are used for each. This can give information about localization of proteins of interest to specific organelles and trafficking of proteins through the cell.

Widefield and confocal microscopy are excellent tools to investigate a protein's localization and trafficking through the cell. In particular, live cell widefield or confocal microscopy is particularly useful for examining these phenomena. Live cell imaging may be used as an initial tool to determine the validity of suspected protein binding partners and trafficking events or to determine the timing of suspected interactions upon addition of a stimulus.

However, additional tools are required to accurately investigate protein–protein interactions. Due to Ernst Abbe's diffraction limit of light given by:

$$d_{x,y} = \lambda/2\mathrm{NA} \qquad (1)$$

the resolution ($d_{x,y}$) of optical microscopes depends on the excitation light wavelength (λ) and the numerical aperture (NA) of the objective. To give an example, when imaging a green fluorophore with 488 nm excitation light through a $100\times$ oil objective with $\mathrm{NA} = 1.4$, the resolution $d_{x,y}$ equals $488\ \mathrm{nm}/(2*1.4) = 174$ nm. On the molecular scale, this is at least one order of magnitude larger than the size of most proteins. Thus, even if two proteins co-localize in diffraction-limited microscopy, we cannot distinguish between true interaction and coexistence in the same diffraction-limited volume. Therefore, advanced microscopy techniques that provide higher-resolution information are required to identify real protein–protein interactions.

One technique for the detection of true protein–protein interactions using standard diffraction-limited fluorescence microscopy is the proximity ligation assay (PLA) [7, 8]. PLA targets two proteins that are suspected to interact (e.g., MIF and NLRP3 [9]) using specific, unlabeled primary antibodies. Importantly, both primary antibodies must be raised in different species (for example, antibody against MIF raised in goat, antibody against NLRP3 raised in rabbit [9]). These primary antibodies are then detected by PLA probes which are secondary, anti-species antibodies linked to unique oligonucleotides (in this example anti-goat probe for MIF and anti-rabbit probe for NLRP3). If the PLA probes are in close proximity to each other (40 nm or less which is indicative of true protein–protein interactions), a subsequently added short connector oligomer can connect the DNA strands. This oligomer is amplified using rolling circle amplification (RCA), a process comparable to a polymerase chain reaction (PCR). The amplified DNA copies can then be specifically hybridized with fluorescently labeled oligonucleotides and detected using fluorescence microscopy.

Another technique commonly used for detecting protein–protein interactions is fluorescence lifetime imaging microscopy (FLIM). In contrast to PLA, FLIM does not require special sample preparation. Instead, special imaging instrumentation is required to perform FLIM. FLIM measures the fluorescence lifetime of a fluorophore within each image pixel. The fluorescence lifetime is defined as the exponential decay rate of fluorescence intensity after excitation with a short laser pulse [10, 11] (for example FLIM images, *see* Fig. 1). A fluorophore's lifetime can be affected by changes in environmental parameters such as the local pH or ion concentration [12]. As such, it can be used to probe for various environmental parameters. Its main application in biological research, however, is the detection of biomolecular interactions. Detection of protein–protein interactions using FLIM is based on fluorescence resonance energy transfer (FRET) [12, 13]. Here, one donor fluorophore and one acceptor fluorophore are used to label two different proteins of interest. If the two fluorophores are within 10 nm proximity to each other, indicating protein–protein interaction, energy can be transferred from the donor fluorophore to the acceptor fluorophore. Energy transfer is a quenching process that affects the excited state of the donor fluorophore by opening additional relaxation pathways. As a consequence, the fluorescence lifetime of the donor fluorophore shortens when FRET occurs. The energy transfer efficiency (E_{FRET}) changes as the inverse sixth power of the distance (r) between donor and acceptor fluorophore:

$$E_{\text{FRET}} = 1/\left[1 + (r/R_0)^6\right] \qquad (2)$$

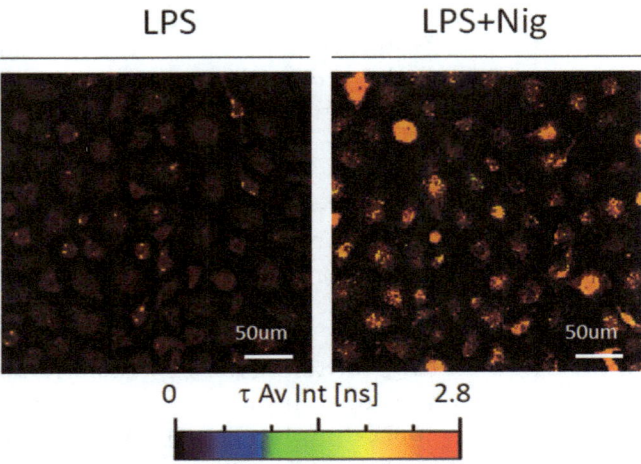

Fig. 1 FLIM imaging of MIF interaction with Vimentin. The fluorescence lifetime image of MIF labeled with Alexa488 (FRET donor) is shown. Vimentin is labeled with Alexa568 (FRET acceptor, not shown). BMMs were treated with lipopolysaccharide (LPS) or LPS + Nigericin. Donor lifetime decrease upon Nigericin treatment indicates increased interaction between MIF and Vimentin

with R_0 being the Foerster radius which is defined as the distance in which FRET efficiency is 50% for any given donor–acceptor fluorophore pair. The energy transfer efficiency can be calculated from the fluorescence lifetime of the donor only (τ_D) and the fluorescence lifetime of the donor in presence of an acceptor (τ_{DA}) using:

$$E_{\text{FRET}} = 1 - (\tau_{DA}/\tau_D) \tag{3}$$

Sample preparation and labeling techniques for FLIM-FRET can be the same as for standard fluorescence microscopy as long as the fluorescence emission spectrum of the donor fluorophore overlaps appropriately with the fluorescence excitation spectrum of the acceptor fluorophore. Commonly used, well-established donor–acceptor fluorophore pairs are, for example, cyan fluorescent protein–yellow fluorescent protein (CFP-YFP) and green fluorescent protein–red fluorescent protein (GFP-RFP) [14].

As FLIM-FRET utilizes genetically encoded fluorophores, it can also be combined with live cell imaging. Live cell FLIM-FRET can be used to study the timing of protein–protein interactions with greater accuracy than widefield or confocal live cell imaging. However, the drawback of live cell FLIM-FRET is the length of time required for data acquisition, meaning often it cannot be applied to interactions which occur rapidly upon introduction of a stimulus, or for very short lived protein–protein interactions.

PLA and FLIM-FRET are complementary techniques that can be used in conjunction to verify each technique's results. If two target proteins interact but the fluorophores cannot approximate closer than 10 nm due to size or orientation of the target proteins,

FLIM-FRET can result in false-negative results due to its inverse sixth power dependence on the fluorophore distance. This problem can be overcome with PLA. However, as with FLIM-FRET, successful PLA requires the availability of well-working, very specific primary antibodies as nonspecific antibody binding can lead to false positive signals.

Despite providing resolution information in the nanometer range, FLIM-FRET is generally not considered a super-resolution technique as the resulting FLIM images remain diffraction limited. Other techniques, such as stochastic optical reconstruction microscopy (STORM) [15, 16] and stimulated emission depletion (STED) microscopy [17, 18], are genuine super-resolution techniques that break the diffraction limit of light and provide images with up to two orders of magnitude higher resolution than diffraction-limited optical microscopy techniques.

STORM is a single molecule localization microscopy (SMLM) technique where fluorescent dye molecules are localized and mapped with nanometer precision [19, 20]. A single fluorophore appears as a diffraction-limited spot of which the center, and therefore the exact localization of the molecule, can be mathematically calculated using Gaussian fitting algorithms. The localization precision depends on the number of photons a single dye molecule can emit during the exposure time of the camera. The more photons, the better the signal-to-noise ratio and the more precise the mathematical algorithm can calculate the exact localization of the molecule. By imaging and mapping hundreds of thousands of single dye molecules, STORM provides a point cloud of 2D (or 3D) localizations of individual molecules. These points can subsequently be reconstructed into a super-resolution image for data visualization.

STED on the other hand is based on confocal imaging, with the addition of high-power Gaussian laser beams which are transformed into a donut shape and overlaid with the confocal excitation lasers [17] (for example, STED images, *see* Fig. 2). This donut-shaped beam induces stimulated (longer wavelength) emission of the outer ring area of the confocal excitation leaving only the center to fluoresce. The resolution of this technique depends on the STED laser power as higher laser powers will reduce the size of the zero-intensity center of the STED donut and consequently reduce the effective focal spot size.

The choice between STED and SMLM will mostly depend on the equipment available. STED imaging and image analysis is generally easier to implement compared to SMLM imaging, especially for multicolor samples. SMLM requires specific sample preparation and the use of precise imaging buffers. For multicolor SMLM, exchange of or compromises in imaging buffer composition are required and may compromise the precision of the technique. Extended data acquisition times for SMLM also necessitates the addition of fiducial markers for sample drift and chromatic

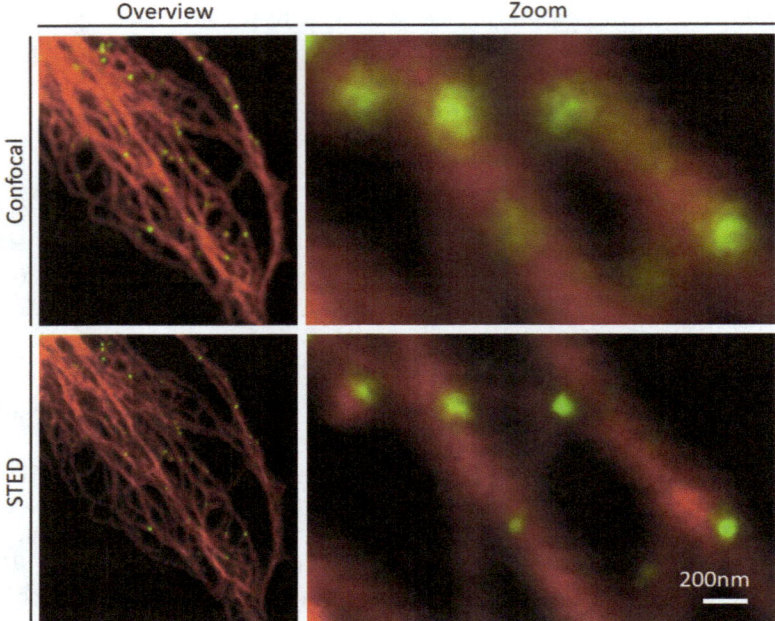

Fig. 2 STED super-resolution imaging of vimentin (red) and MIF interaction partner NLRP3 (green) in wild-type BMMs treated with LPS and nigericin. Redistributed under a Creative Commons License (https:/creativecommons. org/licenses/by/4.0/) from Lang et al. [9]. No modifications were made to the figure

aberration correction. STED microscopy does not require imaging buffers and sample drift will be negligible as image acquisition takes about the same time as standard confocal microscopy. SMLM however can provide higher resolution data, depending on which SMLM technique is chosen, and enables a range of localization-based data analysis options that cannot be performed on image-based microscopy data.

2 Materials

2.1 Cell Culture

1. Bone Marrow-Derived macrophages (BMM).
2. Cell Culture Medium (DMEM) supplemented with 10% FCS (and antibiotics if desired).
3. Phosphate buffered saline (PBS).
4. Trypsin.
5. Sarstedt 8-well chamber on cover glass II (or equivalent) or #1.5 (0.17 mm) glass coverslips suitable for high-resolution and super-resolution microscopy (*see* **Note 1**).

2.2 Transfection of Fluorescently Tagged DNA Expression Vectors

1. DNA expression vectors for proteins of interest linked to fluorescent molecule of choice.
2. Serum-free medium.
3. Transfection reagent (e.g., Lipofectamine 2000, Sigma-Aldrich) (*see* **Note 2**).

2.3 Fixation

1. 100% ice-cold methanol.
2. 4% paraformaldehyde (PFA) in PBS.
3. 2% PFA + 0.02% glutaraldehyde (GA) in PBS.
4. 1% sodium borohydride ($NaBH_4$) in PBS.
5. PBS for washing (*see* **Note 3**).

2.4 Staining for PLA

1. 1% Triton-X 100 in Tris-buffered saline (TBS)
2. PLA reagents (e.g., Duolink® PLA Technology, Sigma-Aldrich).
3. Primary, unconjugated antibody against MIF.
4. Primary, unconjugated antibody against potential interaction partner (raised in different species than MIF antibody).
5. Hoechst/DAPI in TBS.
6. Mounting media.
7. Humidity chamber.

2.5 Staining for FLIM-FRET, STORM, or STED

1. 3% Bovine Serum Albumin (BSA) in PBS.
2. Anti-MIF antibody (conjugated with fluorophores or unconjugated) in PBS with 0.01% Triton-X 100.
3. Antibodies to potential interaction partners (conjugated with fluorophores or unconjugated) in PBS with 0.01% Triton-X 100.
4. Secondary antibodies conjugated to suitable fluorophores (if primary antibodies were unconjugated) in PBS.
5. Mounting media (not required for STORM).

2.6 Confocal Microscopy (Fixed Cell Imaging, Live Cell Imaging, and PLA) Equipment

1. An inverted confocal microscope with suitable laser lines (e.g., 405 nm, 488 nm, 561 nm, and 638 nm) and PMT, GaAsP, or Hybrid confocal detectors.
2. Low magnification air objectives ($10\times$ and $20\times$) are useful to find areas of interest, a high-magnification oil-immersion objective ($60\times$–$100\times$) is recommended for imaging fixed cells, a high-magnification water-immersion objective ($60\times$–$100\times$) for live cell experiments.
3. For live cell imaging the microscope should be fitted with an incubation chamber, which should be warmed and stabilized at

37 °C prior to imaging. Supply of humidified 5% CO_2 to the incubation chamber is also strongly recommended.

4. PLA can be performed on any confocal microscope, thus equipment requirements are as described in 1 and 2.

2.7 FLIM-FRET Microscopy Equipment

1. An inverted confocal microscope additionally equipped with pulsed lasers for FLIM-FRET excitation (438 nm for CFP, 488 nm for GFP).

2. Specialized photon counting detectors and a time correlated single photon counting (TCSPC) card in the data acquisition computer (can be sourced from PicoQuant or Becker&Hickl).

3. A high-capacity data acquisition and analysis computer with at least 32 Gb of RAM, 3 GHz quad-core processor, and 2 Tb of hard-drive storage space is recommended.

2.8 SMLM (Here: STORM) Microscopy Equipment

1. An inverted fluorescence widefield microscope with 405 nm, 561 nm, and 638 nm lasers, capable of performing Total Internal Reflection Fluorescence (TIRF) microscopy. The imaging lasers should have high output power (638 nm: 50 mW, 561 nm: 150 mW). The 405 nm laser serves for fluorophore (re-)activation only. A single-molecule sensitive, cooled EMCCD or sCMOS camera is required with quantum efficiency above 80%. The total magnification of the system should be adjusted so that one camera pixel records 100 nm × 100 nm of the sample.

2. An oil-immersion objective suitable for TIRF illumination (NA = 1.4 or higher) and high magnification (60×–100×) is recommended.

3. A high-capacity data acquisition and analysis computer with at least 32 Gb of RAM, 3 GHz quad-core processor, and 2 Tb of hard-drive storage space is recommended. The computer should enable image acquisition at full camera speed.

2.9 STED Microscopy Equipment

1. An inverted STED microscope with pulsed imaging lasers (488 nm, 561 nm and 638 nm) and high-power STED lasers (775 nm and 595 nm: 1 W).

2. High-magnification oil- and water-immersion objectives (60×–100×) are recommended.

3 Methods

3.1 Cell Culture

General cell culture steps have been described in previous chapters. For microscopy sample preparation, start with confluent BMMs grown in a cell culture dish of your choice.

1. Remove growth medium from cells.

2. Add appropriate amount of Trypsin to cover the cell monolayer.

3. Place cells in 37 °C incubator for ~2 min. The incubation time required will depend on factors such as cell type, confluency, or Trypsin efficacy. Cells should start to detach from the dish before progressing to the next step. Cell detachment can be monitored using an inverted bright-field microscope.

4. Resuspend trypsinized cells in DMEM with 10% FCS.

5. Add appropriate volume of the cell suspension into wells of a Sarstedt chamber or into a cell culture dish containing glass coverslips.

6. Add DMEM supplemented with 10% FCS to top up each well/dish.

7. Place dish in 37 °C incubator overnight or until cells have reattached to the dish.

8. If necessary, stimulate the cells to induce the protein–protein interaction of interest. The duration time for stimulation as well as the stimulus itself will have to be optimized as they depend on the protein–protein interaction of interest.

3.2 Transfection of Fluorescently Tagged DNA Expression Vectors

Precise volumes and incubation times for transfections will be dependent upon the culture dish used and transfection reagent selected. Here we present a general transfection protocol; however all transfection reagents have slightly different protocols and manufacturer's instructions should be consulted before performing transfections. Optimization may be required for your cell line or DNA vector to achieve ideal transfection efficiency (*see* **Note 4**).

1. Begin with BMMs at approximately 60% confluent, preferably in a 6-well dish.

2. For each well to be transfected, combine 100 μL of serum-free medium with 5 μL of transfection reagent.

3. Incubate at room temperature for 5 min.

4. Add 1 mg total of your DNA expression vector (volume will vary depending on your vector concentration).

5. Add the transfection reagent/DNA mix to your well dropwise and gently swirl the dish to mix.

6. Place dish in 37 °C incubator overnight or until fluorescence is observed.

3.3 Preparing Cells for Live Cell Confocal Imaging

For live cell imaging, it is recommended that dishes specifically designed for optical imaging be used. Examples include coverslip thickness glass-bottomed chamber slides (available from Sarstedt or ibidi) or imaging optimized 35 mm dishes (such as the μ-dish available from ibidi).

1. When cells are expressing appropriate fluorescence following transfection with DNA expression vectors, remove growth medium.

2. Add appropriate amount of Trypsin to cover the cell monolayer.

3. Place cells in 37 °C incubator for ~2 min. The incubation time required will depend on factors such as cell type, confluency, or Trypsin efficacy. Cells should start to detach from the dish before progressing to the next step. Cell detachment can be monitored using an inverted bright-field microscope.

4. Resuspend trypsinized cells in DMEM with 10% FCS.

5. Add appropriate volume of the cell suspension into wells of your optical imaging dish or chamber slide.

6. Add DMEM supplemented with 10% FCS to top up each well/dish.

7. Place dish in 37 °C incubator overnight or until cells have reattached to the dish.

8. If necessary, incubate cell-permeable dyes on cells prior to imaging to label any organelles of interest. The incubation times for cell-permeable dyes prior to imaging will have to be optimized and will be dependent on the dye used and cell line. Some dyes must be removed before imaging to avoid oversaturation or mis-localization during the imaging time course. Others can remain on cells for the duration of imaging.

9. If necessary, stimulate the cells to induce the protein–protein interaction of interest. Timing for the addition of stimulus prior to imaging will be dependent on the protein–protein interaction and the speed at which the stimulus induces the interaction. If live cell imaging is being carried out to determine the timing of the interaction upon stimulus, it is recommended that stimulus be added immediately prior to imaging to allow any early effects to be observed.

For long-term live cell imaging *see* **Notes 5** and **6**.

3.4 Fixation

The fixation method to be used will depend primarily on the antibodies chosen as different fixation techniques will affect the target epitopes differently and might hinder or prevent antibody binding. Other considerations are the elected imaging technique (e.g., PFA-only fixation might not fully preserve subcellular fine structures such as microtubules, as a result these structures can appear discontinuous/damaged in super-resolution images) as well as the imaging target (e.g., methanol fixation retains cytoskeletal structures such as actin and microtubules well but is inappropriate for membrane proteins). Therefore, fixation protocols might have to be established and/or optimized for each application. The following protocols are guidelines only and might need to be adapted.

3.4.1 Methanol Fixation

1. Remove medium from cells.
2. Cover cells with 100% ice-cold methanol and incubate for 15 min at $-20\ ^\circ$C.
3. Remove methanol and wash three times with PBS.

3.4.2 PFA Fixation

1. Remove medium from cells.
2. Cover cells with 4% PFA in PBS and incubate for 15 min at room temperature.
3. Remove PFA and wash three times with PBS.

3.4.3 PFA-GA Fixation

1. Remove medium from cells.
2. Cover cells with 2% PFA + 0.02% GA in PBS and incubate for 15 min at room temperature.
3. Remove fixative and wash three times with PBS.
4. It may be necessary to quench free aldehyde fluorescence with sodium borohydride (1% $NaBH_4$ for 20 min); however, if GA is used at very low concentrations, quenching is usually not necessary.

3.5 Staining of Fixed Cells for FLIM-FRET, STORM, or STED

The following protocols for staining are guidelines only. Incubation times, temperatures, buffers, and dilution factors for antibodies will have to be optimized for your antibodies and applications (*see* **Note 7**). Due to the high-/super-resolution information sought, these techniques are highly susceptible to inappropriate sample preparation and optimal sample preparation procedures are crucial (*see* **Notes 8** and **9**)

1. Block cells with 3% BSA in PBS for 30 min.
2. Dilute your preferred primary antibodies in PBS with 0.1% Triton X-100.
3. Remove BSA and incubate overnight at 4 $^\circ$C with primary antibody solution.
4. Dilute your preferred secondary antibodies in PBS.
5. Remove primary antibody solution from your sample and wash three times with PBS.
6. Incubate overnight at 4 $^\circ$C with secondary antibody solution.
7. Remove secondary antibody solution from your sample and wash three times with PBS.

8. If required, add Hoechst/DAPI to your sample at 1:10,000 dilution and incubate for 5 min. Wash thoroughly with PBS.

9. For widefield, confocal, or STED microscopy mount the sample in mounting media. Store the sample in PBS at 4 °C until imaging for STORM.

3.6 Staining of Fixed Cells for PLA

For a successful PLA, it is of paramount importance that proper antibodies and matching reagents are chosen. Except for the primary antibodies, all other reagents necessary for PLA can be purchased as part of a complete PLA kit or as single reagents (e.g., Duolink® PLA Technology, Sigma-Aldrich); however, the right kit/reagents need to be selected based on the primary antibodies used. Where possible, primary antibodies suitable for immunohistochemistry (IHC) or immunofluorescence (IF) should be used as the protocol and the conditions for PLA are very similar to those for IHC/IF (*see* **Notes 7** and **10**).

1. After fixation, let your samples dry at room temperature.

2. Permeabilize cells with 1% Triton-X 100 in TBS for 15 min at room temperature. From this step onward, all incubations need to be performed in a humidity chamber. Make sure that at no point throughout the PLA protocol the cells get dry.

3. Wash twice with TBS.

4. Block for 30 min at room temperature with either the blocking reagent that is recommended for the primary antibody pairs used or the blocking reagent that comes with the PLA kit purchased.

5. Dilute first primary antibody (specific to MIF) in antibody diluent.

6. Remove blocking solution and incubate cells with first primary antibody for 1 h at room temperature. Alternatively incubate overnight at 4 °C.

7. Wash twice with TBS.

8. Dilute second primary antibody (specific to MIF interaction partner) in antibody diluent and incubate on cells for 1 h at room temperature. Alternatively, incubate overnight at 4 °C.

9. Wash twice with TBS.

10. From here onward, steps might be slightly different depending on the PLA reagents purchased. Thus, check the data sheet accompanying your PLA reagents for all following steps.

11. Dilute PLA probes and incubate on cells for 1 h at 37 °C.

12. Wash twice with TBS.

13. Prepare ligation mix and incubate on cells for 30 min at 37 °C.

14. Wash twice with TBS.

15. Prepare amplification mix and incubate on cells for 100 min at 37 °C. From this step onward all incubations should be performed in the dark as the reagents are light sensitive.

16. Wash twice with TBS.

17. Add Hoechst/DAPI to your sample at 1:10,000 dilution and incubate for 5 min. Wash thoroughly with TBS.

18. Add mounting media and keep in the dark at 4 °C until imaging.

3.7 Confocal Microscope Setup and Data Acquisition

1. Place the sample on the microscope, examine the sample to find a region of interest.

2. Switch to "saturation indicator" color scale.

3. Adjust laser intensity and detector gain for each channel so that only a few pixels in the final image are saturated. This ensures that the full dynamic range of the detectors is used without oversaturating the images for image analysis.

4. Record confocal images of each channel.

3.8 Live Cell Confocal Microscopy Setup and Data Acquisition

1. If not kept heated constantly, the incubation chamber and CO_2 delivery system on the microscope should be switched on several hours prior to imaging to allow all microscope components to heat and stabilize.

2. Place the sample on the microscope and examine to find a region of interest.

3. Switch to the saturation indicator and adjust laser intensity and detector gain for each channel so that only a few pixels in the final image are saturated.

4. Set image coordinates for multiple fields of view in the acquisition software if required.

5. Set imaging interval and total imaging time in the acquisition software.

6. It is recommended to recheck the focus of each imaging position prior to beginning your time course to ensure no focal drift has occurred.

7. Begin image acquisition.

3.9 FLIM-FRET Setup and Data Acquisition

The fluorescence lifetime depends on many environmental parameters. Therefore both adequate environmental control and appropriate experimental controls are crucial for successful FLIM-FRET experiments (*see* **Notes 11** and **12**).

1. The repetition rate of the excitation laser must be set appropriately for the dye to be imaged. The fluorescence decay must fit between two laser pulses and background level must be reached before the next excitation pulse. For most commonly used

fluorophores (e.g., autofluorescent proteins like GFP and YFP) a repetition rate of 40 MHz is appropriate. For dyes of unknown lifetime, the appropriate repetition rate must be determined prior to FLIM imaging (*see* **Note 13**).

2. At the start or end of each experiment day, record the instrument response function (IRF) of the FLIM setup using a fluorescent dye of the same color as the donor fluorophore in a saturated potassium iodide solution. Iodide will quench the fluorophore and thus shorten its lifetime to femtoseconds, three orders of magnitude less than the pulse width of the laser (for laser diodes generally in the picosecond range). The decay curve of the quenched fluorophore will therefore approximate the shortest possible process that can be measured with the given equipment and is called IRF. The observed decay in a FLIM measurement is a convolution of the "real" decay with the IRF. To get accurate information on fluorophore lifetimes, recorded lifetimes will need to be de-convolved with the IRF.

3. When performing FLIM-FRET experiments in live cells, a microscope with incubation chamber is required. Turn on heat and CO_2 supply of the microscope's incubation chamber to 37 °C and 5% CO_2 and let equilibrate for 1–2 h. Only when the incubation chamber has reached both set points and provides a stable environment for the cells, place the live cell sample on the microscope and start imaging. Live cell imaging will be performed on cells transfected with fluorescently tagged DNA expression vectors (*see* Subheadings 3.2 and 3.3). When performing FLIM-FRET experiments in fixed cells, **step 2** can be omitted. Fixed cell imaging will be performed on cells stained with appropriate antibodies (*see* Subheadings 3.4 and 3.5).

4. Record confocal images of donor and acceptor fluorophores to be confident that both are present in the area of interest before continuing on to FLIM imaging.

5. Switch to FLIM configuration using pulsed lasers, FLIM detectors, and FLIM acquisition software. Make sure that the standard confocal (continuous wave) lasers are switched off.

6. Record FLIM image at count rates appropriate to the repetition rate of the pulsed laser, recommended is <5%, to prevent pile-up effect (*see* **Note 14**)

3.10 Super-resolution Microscope Setup and Data Acquisition

3.10.1 STORM Super-resolution	1. Add imaging buffer to the well to be imaged. Imaging buffer will need to be optimized for each dye combination and target molecules. For 2-color STORM imaging with Alexa568 and Alexa647 we recommend 50 mM Cysteamine in PBS at pH = 9.0.

2. Place the sample on the microscope, navigate around to find a region of interest and record a widefield image of each channel at low (<5 mW) laser powers.

3. For STORM imaging, start with Alexa647. Set camera exposure time to 20 ms, EMCCD gain to 50, and increase the laser power to ~50 mW. Wait for the dyes to enter the dark state (pumping phase) until you can see individual dye molecules switch on and off (blinking). Then start recording a time series of 10,000 images.

4. Next, change microscope settings to the Alexa568 channel. Camera settings are the same as for Alexa647. Increase the laser power to ~70 mW and enter the pumping phase. When you can see individual dye molecules switch on and off, start recording a time series of 10,000 images.

5. Load each 10,000 frames time series individually into Rapid-STORM or other appropriate STORM software to detect molecule localizations and reconstruct a super-resolution image (*see* **Note 15**).

3.10.2   STED Super-resolution

1. Place the sample on the microscope, navigate around to find a region of interest, and record a confocal image of each channel.

2. Enable the STED laser and record corresponding STED images. Resolution of STED images will depend on the STED laser power applied to the sample. If the STED laser power is set too high, the sample will undergo photo-bleaching during STED image acquisition and a super-resolution image cannot be obtained. Therefore, upon the first instance of imaging a specific sample or antibody with STED, it is recommended to start with low STED power and increase laser power until optimal resolution is achieved.

4   Notes

1. Here we have focused on our representative BMM cell model. Growth medium and reagents may vary depending on cell type used.

2. Details for cloning DNA expression vectors are not covered in the scope of this article. Empty DNA expression vectors for your fluorescent tag of choice (also known as vector backbones) may be purchased from companies such as AddGene

or Clonetech. Your protein of interest can be cloned into the fluorescent backbone by PCR amplification of your target DNA followed by digest of your DNA backbone with restriction enzymes and insertion of your target DNA into the resulting gap.

3. Which fixation reagents (1–4) are required depends on the fixation protocol (*see* Subheading 2.4).

4. Interference with endogenous levels of proteins by transfection with DNA expression vectors can cause so-called overexpression artifacts such as protein mis-localization or abnormal subcellular morphology [21]. Similarly, the addition of a fluorescent molecule such as GFP can affect a protein's normal behavior; the degree of such influence can range from mild (e.g., slowing down protein trafficking due to its increase in size) to severe (e.g., complete blockage of the protein's usual function). Therefore, it is good practice to use labeling of endogenous proteins with antibodies in fixed cells to validate localization of fluorescently tagged proteins in live cells before progressing with protein trafficking experiments.

   In cases of severe effects from overexpression a "knockdown rescue" approach can be utilized where the endogenous protein levels are first reduced by introduction of small interfering RNA (siRNA), followed by re-expression with the fluorescently tagged DNA expression vector.

5. When imaging over long periods of time it is common to experience drift of the sample in the z-plane, which results in a loss of focus over time. This can occur due to changes in heat (which causes expansion or shrinking of metal components of the microscope), loss of oil on high-magnification objectives, or shifting of the dish on the stage. To overcome the issue of focal drift it is important that the incubation chamber on the microscope be heated and stable at 37 °C well before imaging commences. It is also useful to allow the sample to equilibrate in the incubation chamber on the microscope for 30 min to 1 h before setting up image acquisition. This allows the culture vessel to settle on the microscope stage and the sample and culture dish to equilibrate to the same temperature as other components of the imaging system. Ensure the sample is held securely on the stage to help prevent movement over time. It is also recommended for long imaging time courses to use an air- or water-immersion objective rather than an oil-immersion objective to avoid oil loss over time.

6. Although live cell imaging ideally takes place within an incubation chamber supplied with 5% humidified CO_2, incubation setups on microscopes are not a closed system and not as ideal an environment for cell viability as a regular cell culture

incubator. Ensuring that the heat and CO_2 levels are stable can help to maximize the life of cells during imaging. Imaging itself can also induce cell death, in particular on a confocal system where the lasers used for imaging can cause the release of toxic free radicals which ultimately leads to decreased cell viability. To minimize the effects of imaging on cell viability it is recommended that fluorophores be chosen at higher wavelengths, allowing the less toxic near infrared, red, and green lasers to be used for imaging over UV and blue lasers which cause more damage. Imaging at the minimum laser power possible and for shorter times is also advantageous, as is allowing recovery time between imaging intervals, where all lasers are shuttered or switched off, rather than constantly applying laser light to the sample. In extreme cases of toxicity or where very long imaging times are necessary, the introduction of free radical scavengers, such as 6-hydroxy-2,5,7,8-tetramethylchroman2-carboxylic acid (0.1 mM), ascorbate (0.5 mM), and catalase (10 mg/mL), to the cell culture medium prior to imaging can extend the lifetime of the cells. Care should be taken to test whether the introduction of free radical scavengers has any adverse effects on your desired reaction or counter-acts any stimulus being used. This is particularly true for investigating redox reactions.

7. Once the ideal staining and imaging parameters have been established, further optimization might be necessary depending on the protein–protein interaction of interest. For example, some protein–protein interactions do not occur in unstimulated, resting cells but require some sort of cell activation. In that case, the ideal duration of cell activation/stimulation needs to be established, so the cells are fixed after the interaction has been initiated but before the interaction ceases again. Live cell imaging of stimulated cells can be useful in these circumstances to help determine the optimal stimulation time prior to the more complex and costly PLA, FLIM-FRET, or super-resolution imaging experiments being carried out.

8. One of the most critical aspects of super-resolution microscopy is the fixation technique [22, 23]. Fixation artifacts and incomplete preservation of subcellular nanostructures, invisible in diffraction-limited images, become obvious in super-resolved images, thus necessitating adjustments to standard immunofluorescence fixation protocols. For example, in experiments investigating MIF interactions with the cytoskeletal protein Vimentin, we found methanol fixation (*see* Subheading 3.4.1) to be the most appropriate method [9].

9. Another critical aspect of super-resolution microscopy is labeling density [22, 23]. Due to the higher resolution compared to diffraction-limited microscopy techniques, insufficiently

labeled structures can appear discontinued or broken up in super-resolution images. Recommended is a labeling density of at least 2 fluorophores per resolution unit (Nyquist Sampling Theorem). While it is always recommended to check samples on a diffraction-limited microscope before proceeding to a super-resolution microscope, samples optimized for diffraction-limited microscopy are not guaranteed to be suitable for super-resolution microscopy and may need to undergo further optimization for super-resolved imaging. The obvious approach to increase labeling density is increased antibody concentration and extended antibody incubation time. If such approaches fail, other labeling techniques like nanobodies [24, 25] or DNA-PAINT [26, 27] need to be considered.

10. The success of the reaction in a PLA experiment very much depends on the quality and properties of the primary antibodies selected. Often, the ideal reaction conditions for the antibody detecting the first potential interaction partner do not match those for the antibody detecting the second potential interaction partner. Careful and sometimes lengthy optimizations are necessary to get both primary antibodies working under the same reaction conditions. Thus, it is advisable to establish the ideal binding conditions for the primary antibody pair in a conventional IHC or IF experiment (using fluorophore-conjugated secondary antibodies for detection), before attempting to perform a PLA. Similar to general staining optimizations, parameters that can be altered to establish the ideal binding conditions for a given primary antibody pair include fixation method, blocking, primary antibody dilutions and incubation times and temperatures.

    False-positive results on one hand can occur when primary antibodies bind to unspecific target proteins, thus mimicking an interaction that actually did not occur. The same is true when primary antibodies have not been washed off the cells properly. False-negative results on the other hand can result from antibodies not working under the given reaction conditions. Therefore, careful optimization as described above is an absolute requirement before PLA should be performed.

    But even after optimization, it is good practice to always include proper negative, and although not always possible, positive controls to monitor for the accuracy of the PLA reaction. Cells that are deplete of one of the interaction partners or, where a protein–protein interaction requires cell activation, unstimulated cells can serve as negative controls. In addition, technical negative controls (i.e., cells that receive only one of the primary antibodies) should be included. These conditions are not only helpful in detecting false-positive signals but can also be used to determine background signal intensity.

Especially for the detection of previously unknown protein–protein interactions, it is not always possible to include an appropriate positive control. However, if it is already known that an interaction occurs, for example, under specific stimulation conditions and the goal is to determine whether that interaction occurs under different stimulation conditions as well, cells treated to induce the known interaction should be included as a positive control. This can help to exclude false-negative results that can occur after using faulty reagents or PLA reaction conditions that are too stringent.

11. The fluorescence lifetime of a specific dye depends on many environmental parameters, including local pH value, membrane potential, and ion concentration [12, 28, 29]. While this makes FLIM-FRET a powerful tool to analyze those parameters, it can also hinder FLIM-FRET analysis if multiple parameters change between experiments. Especially in biological applications, where individual specimens exhibit intrinsic variations, care must be taken in experimental design and execution to minimize such variations and appropriate control experiments must be performed. Specifically, it is important to keep the pH value the same for all samples when doing FLIM-FRET, i.e., make sure all solutions are adjusted to the same pH and, if possible, use the same stock solutions for all samples. Important controls for FLIM-FRET are donor-only controls (i.e., samples prepared under the same conditions as donor–acceptor samples but without acceptor staining) and unstained samples to assess the tissue-specific autofluorescence and its contribution to the recorded FRET donor lifetime.

12. Another important consideration is autofluorescence contribution to the lifetime decay curve. While autofluorescence might not be obvious in the confocal images, it will still contribute to the recorded fluorescence decay curve and affect the recorded lifetime. Autofluorescence lifetimes derive from a variety of compositions of subcellular components and thus strongly differ between biological specimens and tissues, making autofluorescence contribution to FLIM difficult to compensate for. However, these characteristics also allow for autofluorescence itself to be a valuable source of information or for the differentiation between autofluorescence and the fluorescent label [12].

13. For accurate calculation of fluorescence lifetimes, correct selection of laser repetition rates is crucial. The fluorescence decay must fit between two laser pulses and background level must be reached before the next excitation pulse. If the laser repetition rate is too high and excitation pulses occur too frequently, each decay will bleed into the subsequent decay and distort its

shape, thus hindering accurate data analysis. Too low laser repetition rates on the other hand will not affect data analysis but will unnecessarily lengthen the duration of data acquisition as after each decay background only is recorded until the next laser pulse occurs. Care must be taken not to choose too high repetition rates and if in doubt, e.g., because a fluorophore with unknown lifetime is being recorded for the first time, it is recommended to start with the lowest available laser repetition rate and increase the laser repetition rate as high as possible while still having the decay curve reach noise level before the next excitation pulse arrives.

14. FLIM data can suffer from artifacts if data acquisition was not performed appropriately. The most common artifact is the so-called pile-up effect [10, 11]. Most single photon detectors and TCSPC electronics suffer from dead time, a certain time frame after a photon was registered in which subsequently arriving photons cannot be detected because the system is busy with data processing. Therefore, when the photon count rate is too high, early photons are statistically overrepresented, the resulting fluorescence decay histogram will be distorted and the measured lifetimes will appear shortened. To prevent pile-up effect, it is important to acquire FLIM data at count rates appropriate to the repetition rate of the pulsed laser. Recommended is <5%, to prevent pile-up effect. Example: If the repetition rate of the laser is set to 40 MHz, the count rate should not exceed 2 MHz. At a pixel dwell time of 40 µs this corresponds to a maximum of $4,000,000[1/s]*2E-5[s] = 80$ photons per pixel. To adjust the photon count rate, adjust the laser power using a physical attenuator (such as neutral density filters), not in the software as this would change the IRF.

15. After successful data acquisition, the next challenge is to analyze the data appropriately. Super-resolution data analysis, specifically localization microscopy (here: STORM) data analysis, is challenging [30, 31], not only due to the comparably large data sets (one raw image series can be several Gb in size) but also due to the novelty of the technique and the vastly different type of data it provides us with. In contrast to most other imaging techniques, STORM does not provide images as we are used to, instead, we obtain a point cloud of 2D (or 3D) localizations of individual dye molecules. These points can be reconstructed into a so-called "pointillist" image that resembles what we are used to from other imaging techniques. However, STORM data analysis should be performed on the original point cloud and thus necessitates the development of novel data analysis algorithms.

References

1. Combs CA (2010) Fluorescence microscopy: a concise guide to current imaging methods. Curr Protoc Neurosci Chapter 2:Unit2.1. https://doi.org/10.1002/0471142301.ns0201s50

2. Combs CA, Shroff H (2017) Fluorescence microscopy: a concise guide to current imaging methods. Curr Protoc Neurosci 79:2.1.1–2.1.25. https://doi.org/10.1002/cpns.29

3. Pawley JB (2006) Handbook of biological confocal microscopy. Springer, New York

4. Hibbs AR (2004) Confocal microscopy for Biologists. Springer, New York

5. Calandra T, Roger T (2003) Macrophage migration inhibitory factor: a regulator of innate immunity. Nat Rev Immunol 3 (10):791–800. https://doi.org/10.1038/nri1200

6. Kim TK, Eberwine JH (2010) Mammalian cell transfection: the present and the future. Anal Bioanal Chem 397(8):3173–3178. https://doi.org/10.1007/s00216-010-3821-6

7. Fredriksson S, Gullberg M, Jarvius J, Olsson C, Pietras K, Gústafsdóttir SM, Östman A, Landegren U (2002) Protein detection using proximity-dependent DNA ligation assays. Nat Biotechnol 20:473. https://doi.org/10.1038/nbt0502-473

8. Olink - Precision Proteomics for Life. https://www.olink.com/

9. Lang T, Lee JPW, Elgass K, Pinar AA, Tate MD, Aitken EH, Fan H, Creed SJ, Deen NS, Traore DAK (2018) Macrophage migration inhibitory factor is required for NLRP3 inflammasome activation. Nat Commun 9(1):2223. https://doi.org/10.1038/s41467-018-04581-2

10. Lakowicz JR (2006) Principles of Fluorescence Spectroscopy. Springer, New York

11. Becker W (2017) The bh TCSPC handbook. Becker and Hickl GmbH, Berlin

12. Becker W (2012) Fluorescence lifetime imaging – techniques and applications. J Microsc 247(2):119–136. https://doi.org/10.1111/j.1365-2818.2012.03618.x

13. Bastiaens PI, Squire A (1999) Fluorescence lifetime imaging microscopy: spatial resolution of biochemical processes in the cell. Trends Cell Biol 9(2):48–52

14. Bajar BT, Wang ES, Zhang S, Lin MZ, Chu J (2016) A guide to fluorescent protein FRET Pairs. Sensors (Basel, Switzerland) 16(9). https://doi.org/10.3390/s16091488

15. Rust MJ, Bates M, Zhuang X (2006) Stochastic optical reconstruction microscopy (STORM) provides sub-diffraction-limit image resolution. Nat Methods 3(10):793–795. https://doi.org/10.1038/nmeth929

16. van de Linde S, Loeschberger A, Klein T, Heidbreder M, Wolter S, Heilemann M, Sauer M (2011) Direct stochastic optical reconstruction microscopy with standard fluorescent probes. Nat Protoc 6(7):991–1009. https://doi.org/10.1038/nprot.2011.336

17. Hell SW, Wichmann J (1994) Breaking the diffraction resolution limit by stimulated emission: stimulated-emission-depletion fluorescence microscopy. Opt Lett 19(11):780–782. https://doi.org/10.1364/OL.19.000780

18. Hein B, Willig KI, Hell SW (2008) Stimulated emission depletion (STED) nanoscopy of a fluorescent protein-labeled organelle inside a living cell. Proc Natl Acad Sci U S A 105 (38):14271–14276. https://doi.org/10.1073/pnas.0807705105

19. Sauer M, Heilemann M (2017) Single-molecule localization microscopy in eukaryotes. Chem Rev 117(11):7478–7509. https://doi.org/10.1021/acs.chemrev.6b00667

20. Allen JR, Ross ST, Davidson MW (2013) Single molecule localization microscopy for super-resolution. J Opt 15:094001. https://doi.org/10.1088/2040-8978/15/9/094001

21. Gibson TJ, Seiler M, Veitia RA (2013) The transience of transient overexpression. Nat Methods 10:715. https://doi.org/10.1038/nmeth.2534

22. Allen JR, Ross ST, Davidson MW (2013) Sample preparation for single molecule localization microscopy. Phys Chem Chem Phys 15 (43):18771–18783. https://doi.org/10.1039/c3cp53719f

23. Whelan DR, Bell TD (2015) Image artifacts in single molecule localization microscopy: why optimization of sample preparation protocols matters. Sci Rep 5:7924. https://doi.org/10.1038/srep07924

24. Pleiner T, Bates M (2015) Nanobodies: site-specific labeling for super-resolution imaging, rapid epitope-mapping and native protein complex isolation. Elife 4:e11349. https://doi.org/10.7554/eLife.11349

25. Traenkle B, Rothbauer U (2017) Under the microscope: single-domain antibodies for live-cell imaging and super-resolution microscopy. Front Immunol 8:1030. https://doi.org/10.3389/fimmu.2017.01030

26. Schnitzbauer J, Strauss MT, Schlichthaerle T, Schueder F, Jungmann R (2017) Super-resolution microscopy with DNA-PAINT. Nat Protoc 12(6):1198–1228. https://doi.org/10.1038/nprot.2017.024

27. Nikic-Spiegel I (2018) Genetic code expansion- and click chemistry-based site-specific protein labeling for intracellular DNA-PAINT imaging. Methods Mol Biol 1728:279–295. https://doi.org/10.1007/978-1-4939-7574-7_18

28. Korczynski J, Wlodarczyk J (2009) Fluorescence lifetime imaging microscopy (FLIM) in biological and medical research. Postepy Biochem 55(4):434–440

29. van Munster EB, Gadella TW (2005) Fluorescence lifetime imaging microscopy (FLIM). Adv Biochem Eng Biotechnol 95:143–175

30. Wheeler A, Henriques R (2017) Standard and super-resolution bioimaging data analysis: a primer. Wiley, Hoboken, NJ

31. Lee A, Tsekouras K, Calderon C, Bustamante C, Presse S (2017) Unraveling the thousand word picture: an introduction to super-resolution data analysis. Chem Rev 117(11):7276–7330. https://doi.org/10.1021/acs.chemrev.6b00729

Chapter 10

Co-Immunoprecipitation of Macrophage Migration Inhibitory Factor

Jawad H. Abidi, James Harris, and Nadia S. Deen

Abstract

Immunoprecipitation is a technique which enables a macromolecule of interest to be isolated from heterogenous mixtures (particularly cell lysates). However, the immunoprecipitation of protein(s) can be challenging, with multiple variations of the basic technique required for successful antigenic pull-down. This depends on the target of interest, cell source, and localization. Here, immunoprecipitation of MIF from mouse and human macrophage cell lysates is described, which is both reliable and replicable, derived from multiple optimization experiments.

Key words MIF, Macrophage migration inhibitory factor, Immunoprecipitation, Co-immunoprecipitation, Flowthrough, Whole cell lysate

1 Introduction

Despite being one of the first immune active molecules discovered (in the 1960s), the full repertoire of binding partners of macrophage migration inhibitory factor (MIF), as well as the specific biochemical pathways it is involved in is not known [1]. At a disease level, MIF is known to be associated with pathology in several inflammatory autoimmune conditions, with genome distribution studies implicating specific gene polymorphisms with disease phenotypes [1, 2]. Additionally, MIF may play a role in amyotrophic lateral sclerosis (ALS) as a protective chaperone protein for Cu/Zn superoxide dismutase (SOD1) [3].

MIF is induced by glucocorticoids and may serve to counter their anti-inflammatory effects through a complex mechanism involving glucocorticoid induced leucine zipper (GILZ) [4]. In addition, MIF is involved in autophagy [5–12] and in the release of several pro-inflammatory cytokines, including IL-1 family cytokines [13]. MIF exists as a homotrimer and exhibits enzymatic activity (specifically, tautomerase and thiol-protein oxidoreductase

James Harris and Eric F. Morand (eds.), *Macrophage Migration Inhibitory Factor: Methods and Protocols*,
Methods in Molecular Biology, vol. 2080, https://doi.org/10.1007/978-1-4939-9936-1_10,
© Springer Science+Business Media, LLC, part of Springer Nature 2020

activity) in vivo [14–23]. However, physiological substrates of clinical significance have yet to be identified.

Immunoprecipitation of MIF from mouse and human macrophages (and other cells) is an important technique, allowing for detailed study of the MIF interactome. The basic principle of immunoprecipitation centers around isolating a protein or ribonucleic acid (RNA) segment from biological samples (typically cell lysates) using an antibody that binds with the antigen of interest. Different forms of immunoprecipitation exist, where each variation aims to separate either a single protein (simple immunoprecipitation or "IP"), protein–protein complexes (protein complex immunoprecipitation or "Co-IP"), histone proteins (chromatin immunoprecipitation or "Ch-IP"), or RNA segment (RNA immunoprecipitation or "R-IP") from a sample of interest.

One of two basic protocols is used for most immunoprecipitation experiments, "direct" or "indirect." The direct method involves the binding of the chosen antibody to either agarose or magnetic microbeads, and then finally adding the protein mixture (which contains the antigen destined for isolation) with the antibody–microbead complex. The indirect method is comparatively different in that the antibody is added directly to the protein mixture, and the microbeads added thereafter once the antigen has bound the antibody. This chapter outlines our protocols for direct immunoprecipitation of MIF from human and mouse macrophages.

2 Materials

2.1 Cell Culture and Preparation of Lysates

1. Complete medium: RPMI-1640 medium supplemented with 10% Fetal Calf Serum (FCS) filtered through a 0.2 μM filter, 50 μg/mL Streptomycin, 50 U/mL Penicillin, and 2 mM L-Glutamate. Stored at 4 °C.

2. Phosphate-buffered saline (PBS): 137 mM NaCl, 10 mM Na_2HPO_4, 1.8 mM KH_2PO_4, 2.7 mM KCl in MilliQ H_2O, adjust to pH 7.4 with HCl. Sterilize by autoclaving or filtration. Store at 4 °C.

3. RIPA Buffer: 1% (v:v) igepal® CA-630, 150 mM Sodium Chloride, 50 mM Tris (pH 8.0), 0.5% (w:v) Sodium Deoxycholate, 0.1% (w:v) Sodium Dodecyl Sulfate (see **Note 1**). Add EDTA/EDTA-free protease inhibitor at concentration recommended by supplier for specific application (see **Note 2**).

4. Cell scrapers.

5. 50 mL tubes.

2.2 Preclearing Lysates and Incubation of Pull-Down Antibody with Dynabeads®

1. Dynabeads® (Protein G) (*see* **Note 3**).
2. Dynabeads® magnet.

2.3 Crosslinking of Dynabeads® Protein G with Pull-Down Antibody

1. Conjugation Solution Buffer: 20 mM HEPES (4-(2-hydro-xyethyl)-1-piperazineethanesulfonic acid). Store at room temperature.
2. Quenching Buffer: 1 M Tris–HCl (pH 7.5). Store at room temperature.
3. 100 mM BS^3 (Bissulfosuccinimidyl suberate). Store at room temperature.
4. 5 mM BS^3 (Bissulfosuccinimidyl suberate). Store at room temperature.
5. PBS with 0.05% Tween-20 (PBS-T): Store at room temperature.

2.4 Immuno-precipitation

1. Wash solution: 150 mM Sodium chloride, 50 mM Tris (pH 8.0). Keep on ice.

2.5 Post Co-IP Sample Processing

1. Neutralizing solution: 1 M Tris (pH 8). Store at room temperature.
2. Eluting solution 1: 1:1 NuPAGE MOPS 2x Sample Buffer and 2× NuPAGE Sample Reducing Agent. Store at room temperature.
3. Eluting solution 2: 0.2 M Glycine (pH 2.5). Store at room temperature.

3 Methods

The following steps are to be performed at room temperature, unless otherwise indicated.

3.1 Cell Culture and Preparation of Lysates

1. Plate 5×10^6–10×10^6 cells and pipette them into 10 cm plates or 75 cm^2 flasks in 10 mL complete medium (*see* **Note 4**). Allow to settle overnight in an incubator at 37 °C.
2. Treat the cells as required or leave them untreated, as per your experimental design.
3. Wash the cells twice in ice-cold PBS to thoroughly remove the media +/− stimulating agents.
4. Add 500 μL of RIPA lysis buffer to the cells, rotating the plates so the lysis buffer coats the cell occupied surface (*see* **Note 5**).

5. Leave the cells on ice for 5 min.

6. Scrape the cell lysates using a cell scrapper and transfer the lysates to Eppendorf tubes.

7. Keep the Eppendorf tubes at 4 °C for 1 h with agitation to complete the lysis.

8. Centrifuge the cell lysates in a microcentrifuge at $\sim 10,000 \times g$ for 10–15 min (*see* **Note 6**).

9. The supernatants may then be used for immunoprecipitation and the pellet (nucleic acid and large organelle debris) can be discarded.

3.2 Preclearing Lysates

1. Pipette 20 μL of Dynabeads® Protein G into separate Eppendorf tubes as per number of lysates.

2. Wash the Dynabeads® Protein G twice with 50 μL RIPA lysis buffer (commercial or self-made), using a magnet to pellet out the Dynabeads® Protein G between washes.

3. Add Dynabeads® Protein G to each of the lysates.

4. Leave the lysates at 4 °C for 1 h with agitation to complete preclearing.

5. Remove and then dispose of the Dynabeads® Protein G from the lysates using a magnet and transfer the lysate to fresh Eppendorf tubes.

6. Draw 30–50 μL from each lysate and pipette them into new Eppendorf tubes for parallel whole cell lysate analysis.

3.3 Crosslinking of Dynabeads® Protein G with Pull-Down Antibody

This step is optional and may also be included after the Dynabeads® Protein G and antibody have been left to incubate together.

1. Add 200 μL Conjugation Buffer to the Antibody-Dynabeads® Protein G and ensure that they are evenly dispersed.

2. Pellet out the Antibody-Dynabeads® Protein G using the magnet and discard the solution.

3. Add 250 μL 5 mM BS^3 (prepared from 100 mM BS^3) and ensure the Antibody-Dynabeads® Protein G are dispersed well. Incubate at room temperature for 30 min with tilting/agitation.

4. Separate the solution from the Antibody-Dynabeads® Protein G using the magnet and remove the supernatant.

5. Add 12.5 μL Quenching Buffer to the Antibody-Dynabeads® Protein G. Incubate at room temperature for 15 min with tilting/agitation.

6. Wash the Antibody-Dynabeads® Protein G three times with 200 μL PBS-T, using the magnet to separate the beads between each rinse.

**3.4 Immuno-
precipitation**

1. Transfer 50 μL of Dynabeads® Protein G to new Eppendorf tubes.

2. Rinse twice in 50 μL RIPA buffer, using the magnet to pellet the Dynabeads® Protein G between each rinse.

3. Add 10 μL of chosen anti-MIF Ab to the Dynabeads® Protein G (1: 50 dilution) and ensure they are evenly distributed throughout. Incubate for 10–15 min with agitation (*see* **Note 7**).

4. Use the magnet to pellet the Antibody-Dynabeads® Protein G complex and remove the supernatant containing excess antibody.

5. Wash Dynabeads® Protein G twice in 50 μL RIPA buffer, using the magnet to pellet them between each rinse.

6. Add the pre-cleared lysates (500 μL) to each Dynabeads® Protein G containing Eppendorf tube and mix well.

7. Incubate at room temperature for 30 min, or at 4 °C for 2 h.

8. Use the magnet to pellet the antigen-antibody-Dynabeads® Protein G complex and pipette the remaining lysate ("flow-through") into new Eppendorf tubes. These may be used in SDS-PAGE or Western blot analysis.

9. Wash the Dynabeads® Protein G 3–5 times with 200 mL of wash solution, pelleting out the Dynabeads® Protein G using the magnet between each individual rinsing (*see* **Note 8**).

3.5 Elution

The specific method of eluting the antigen of interest from the Dynabeads® Protein G will vary depending on the experimental design of the researcher. Here, we detail 2 elution protocols: one for preparing samples for Western blot, the other for mass spectrometry.

*3.5.1 Eluting Samples
for Western Blot Analysis*

1. Add 50–70 μL eluting solution 1 to the Dynabeads® Protein G and allow them to sit for 10 min at room temperature.

2. Heat the samples to 95 °C for 5 min (*see* **Note 9**).

3. Use the magnet to pellet the Dynabeads® Protein G, and pipette the mixture now containing the sample proteins to a new Eppendorf tube.

4. Centrifuge the samples in a microcentrifuge at $\sim 10,000 \times g$ for 5 min. These are ready to analyze by Western blot.

*3.5.2 Eluting Samples
for Mass Spectrometry
Analysis*

1. Add 50 μL 0.2 M Glycine to the Dynabeads® Protein G and allow them to sit for 10 min at room temperature.

2. Heat the samples to 95 °C for 5 min.

3. Use the magnet to pellet the Dynabeads® Protein G, and pipette the mixture now containing the sample proteins to a new Eppendorf tube.

4. Neutralize the pH of the samples using the 1M Tris-base neutralizing solution. Use pH strips to confirm neutralization (*see* **Note 10**).

5. Centrifuge in a microcentrifuge at $\sim 10,000 \times g$ for 5 min.

4   Notes

1. Adding the MilliQ H_2O first, then the remainder of the ingredients allows an easier and more thorough dissolution. Additionally, cutting the end of a micropipette tip prior to pipetting igepal® CA-630 allows for a more accurate and consistent measurement of volume, given the reagent's viscosity.

2. Any remaining protease inhibitor solution can be stored at −20 °C, ready for future use once thawed.

3. Preclearing of lysates with 20 μL Dynabeads® Protein G reduces the amount of nonspecific protein binding to the samples. Co-IP samples without preclearing may contain proteins that were bound directly to the Dynabeads® Protein G and not the antigen of interest. Preclearing decreases this possibility.

4. We have used this protocol with primary murine bone marrow-derived macrophages (BMM), immortalized BMM, and human THP-1 monocytic cells. For non-adherent THP-1 cells, the process of lysis differs. Transfer cells to 50 mL tubes and centrifuge at $\sim 400 \times g$ for 5 min. Resuspend in sterile PBS and repeat. After second wash, discard PBS and add lysis buffer directly to the cell pellet.

5. Commercial RIPA buffer is also available.

6. Centrifugation of the lysates allows for the removal of nonprotein components from the samples, namely nucleic acids, which may interfere with results.

7. Incubating Dynabeads® Protein G with antibody for longer than 15 min has no further positive bearing on immunoprecipitation procedure.

8. Rinsing with wash solution removes any remaining lysate that could interfere with the experimental results. However, overly vigorous rinsing could result in partial loss of the sample. Very gentle pipetting or simply tapping the edge of the Eppendorf to distribute the Dynabeads® Protein G during each wash instead of pipetting helps preserve the sample proteins.

9. Heating at 70 °C for 10 min is equally as effective at elusion as heating at 95 °C for 5 min.

10. The faster the pH of these samples is neutralized after elusion, the better. Keeping the samples in a low pH environment degrades both the binding partners and the target antigen after some time.

References

1. Lang T et al (2015) MIF: Implications in the pathoetiology of systemic lupus erythematosus. Front Immunol 6:577

2. Kim KW, Kim HR (2016) Macrophage migration inhibitory factor: a potential therapeutic target for rheumatoid arthritis. Korean J Intern Med 31(4):634–642

3. Leyton-Jaimes MF, Kahn J, Israelson A (2018) Macrophage migration inhibitory factor: a multifaceted cytokine implicated in multiple neurological diseases. Exp Neurol 301 (Pt B):83–91

4. Fan H et al (2014) Macrophage migration inhibitory factor inhibits the antiinflammatory effects of glucocorticoids via glucocorticoid-induced leucine zipper. Arthritis Rheumatol 66(8):2059–2070

5. Chuang YC et al (2012) Macrophage migration inhibitory factor induces autophagy via reactive oxygen species generation. PLoS One 7(5):e37613

6. Xu X et al (2013) Macrophage migration inhibitory factor plays a permissive role in the maintenance of cardiac contractile function under starvation through regulation of autophagy. Cardiovasc Res 99(3):412–421

7. Xu X et al (2014) Macrophage migration inhibitory factor deletion exacerbates pressure overload-induced cardiac hypertrophy through mitigating autophagy. Hypertension 63 (3):490–499

8. Chen HR et al (2015) Macrophage migration inhibitory factor induces vascular leakage via autophagy. Biol Open 4(2):244–252

9. Lee JP et al (2016) Loss of autophagy enhances MIF/macrophage migration inhibitory factor release by macrophages. Autophagy 12 (6):907–916

10. Xia W, Hou M (2016) Macrophage migration inhibitory factor induces autophagy to resist hypoxia/serum deprivation-induced apoptosis via the AMP-activated protein kinase/mammalian target of rapamycin signaling pathway. Mol Med Rep 13(3):2619–2626

11. Xu S et al (2016) Macrophage migration inhibitory factor enhances autophagy by regulating ROCK1 activity and contributes to the escape of dendritic cell surveillance in glioblastoma. Int J Oncol 49(5):2105–2115

12. Chao CH et al (2018) Macrophage migration inhibitory factor-induced autophagy contributes to thrombin-triggered endothelial hyperpermeability in sepsis. Shock 50(1):103–111

13. Harris J et al (2019) Rediscovering MIF: new tricks for an old cytokine. Trends Immunol. https://doi.org/10.1016/j.it.2019.03.002

14. Rosengren E et al (1997) The macrophage migration inhibitory factor MIF is a phenylpyruvate tautomerase. FEBS Lett 417(1):85–88

15. Lubetsky JB et al (2002) The tautomerase active site of macrophage migration inhibitory factor is a potential target for discovery of novel anti-inflammatory agents. J Biol Chem 277 (28):24976–24982

16. Senter PD et al (2002) Inhibition of macrophage migration inhibitory factor (MIF) tautomerase and biological activities by acetaminophen metabolites. Proc Natl Acad Sci 99(1):144–149

17. Garai J, Lorand T (2009) Macrophage migration inhibitory factor (MIF) tautomerase inhibitors as potential novel anti-inflammatory agents: current developments. Curr Med Chem 16(9):1091–1114

18. Kobold S et al (2014) The macrophage migration inhibitory factor (MIF)-homologue D-dopachrome tautomerase is a therapeutic target in a murine melanoma model. Oncotarget 5(1):103–107

19. Dziedzic P et al (2015) Design, synthesis, and protein crystallography of biaryltriazoles as potent tautomerase inhibitors of macrophage migration inhibitory factor. J Am Chem Soc 137(8):2996–3003

20. Sarkar S et al (2015) Ellagic acid, a dietary polyphenol, inhibits tautomerase activity of human macrophage migration inhibitory factor and its pro-inflammatory responses in human peripheral blood mononuclear cells. J Agric Food Chem 63(20):4988–4998

21. Guo D et al (2016) D-dopachrome tautomerase is over-expressed in pancreatic ductal

adenocarcinoma and acts cooperatively with macrophage migration inhibitory factor to promote cancer growth. Int J Cancer 139 (9):2056–2067

22. Zhang Y et al (2016) Inhibition of macrophage migration inhibitory factor (MIF) tautomerase activity suppresses microglia-mediated inflammatory responses. Clin Exp Pharmacol Physiol 43(11):1134–1144

23. Roger T et al (2017) Plasma levels of macrophage migration inhibitory factor and d-dopachrome tautomerase show a highly specific profile in early life. Front Immunol 8:26

Chapter 11

Concurrent Immunohistochemical Localization and Western Blot Analysis of the MIF Receptor, CD74, in Formalin-Fixed, Paraffin-Embedded Tissue

Amanda Graham and Warren B. Nothnick

Abstract

Archived, formalin-fixed, paraffin-embedded tissue provides a robust resource for assessing protein expression in a variety of complex tissue types. Immunohistochemical localization techniques allow one to identify proteins of interest in the different cell populations which compose these tissues, but quantitative comparison within and between samples is semiquantitative. In contrast, Western blot analysis provides a more quantitative assessment but without the ability to identify the cellular sources of expressed protein. Here we describe a dual approach using human endometrium to assess both the localization and quantitation of the macrophage migration inhibitory factor (MIF) receptor CD74 by immunohistochemical techniques and Western blotting.

Key words Formalin-fixed, paraffin-embedded tissue, Western blot, Immunohistochemistry

1 Introduction

Macrophage migration inhibitory factor (MIF) is a potent proinflammatory cytokine which was first described over 50 years ago in studies of delayed hypersensitivity [1, 2]. MIF signal transduction is initiated by binding with its cell surface receptor, CD74 [3]. Since its initial description, MIF and its receptor have been implicated to play a role in the pathophysiology of cancer [4, 5], pulmonary inflammation [6], and kidney disease [7], as well as in the pathophysiology of endometriosis [8].

Identification of MIF and/or CD74 in these as well as other tissues have utilized archived, formalin-fixed, paraffin-embedded (FFPE) tissue. Using immunohistochemistry (IHC), the cellular sources of MIF/CD74 can be identified in these complex tissues. From a physiology/pathophysiology perspective, identifying the cellular sources provides important mechanistic/signaling information. However, the downside is that IHC is only semiquantitative.

James Harris and Eric F. Morand (eds.), *Macrophage Migration Inhibitory Factor: Methods and Protocols*,
Methods in Molecular Biology, vol. 2080, https://doi.org/10.1007/978-1-4939-9936-1_11,
© Springer Science+Business Media, LLC, part of Springer Nature 2020

Quantitative assessment of protein expression can be accomplished by Western blot analysis, but extraction of protein from complex tissue and subsequent analysis using this approach leads to the inability to discern cellular sources of the protein.

2 Materials

2.1 Tissue Processing

1. 10% Neutral buffered formalin.
2. Ethanol: 100%, 95%, 80%, 70% diluted in water.
3. Xylenes.
4. Paraplast.
5. Tissue embedding cassettes.
6. Leica ASP300 smart tissue processor.
7. Leica EG1150C and H Paraffin Embedding Station.
8. Leica RM2125 Microtome.

2.2 Protein Extraction and Quantification

1. Qproteome FFPE Tissue Kit.
2. Extraction Buffer EXB (*see* **Note 1**).
3. Bovine Serum Albumin, 100 mg/mL stock solution in water. Make dilutions in water to obtain 2 mg/mL, 1 mg/mL, 0.5 mg/mL, 0.25 mg/mL, and 0.125 mg/mL standards to be used for the standard curve.
4. Protein Assay kit (*see* **Note 2**).
5. Plastic cuvette.
6. Spectrophotometer.

2.3 Western Blot

1. 40% Acrylamide/Bis.
2. 1 M Tris–HCl, pH 8.8.
3. 1 M Tris–HCl, pH 6.8.
4. 10% Sodium Dodecyl Sulfate (SDS) solution in water.
5. N,N,N,N'-Tetramethyl-ethylenediaminie (TEMED).
6. 10% Ammonium persulfate (APS) solution in water, freshly prepared.
7. Isopropanol.
8. Mini PROTEAN spacer plates with 1.5 mm spacers.
9. Mini PROTEAN short plates.
10. Mini PROTEAN 1.5 mm combs.
11. Mini PROTEAN Tetra Cell Casting Stand with Clamp Kit.
12. Mini PROTEAN Tetra Electrophoresis system.
13. Mini Trans-Blot Cell.

14. Power supply.

15. Running buffer (10×): 900 mL water, 22.5 g Tris base, 108 g Glycine, 10 g SDS. On a stir plate, ensure all reagents are dissolved and then Qs to 1 L with water.

16. Running buffer (1×): 900 mL water, 100 mL 10× Running buffer.

17. Transfer buffer (10×): 900 mL water, 30.3 g Tris base, 140 g Glycine, 1 g SDS. On a stir plate, ensure all reagents are dissolved and then Qs to 1 L with water. Store at 4 °C.

18. Transfer buffer (1×): 700 mL water, 100 mL 10× Transfer buffer, 200 mL methanol.

19. Sample Buffer: 1 mL 1 M Tris–HCl pH 6.8, 1 mL water, 1.6 mL glycerol, 3.2 mL 10% SDS, 800 μL 2-mercaptoethanol, 400 μL 0.05% bromophenol blue.

20. Protein standards.

21. Pyrex dish.

22. Nitrocellulose membrane.

23. Tris Buffered Saline (TBS, 10×): 800 mL water, 30 g Tris Base, 80 g NaCl, 2 g KCl. On a stir plate, ensure all reagents are dissolved, adjust pH to 7.4, and Qs to 1 L.

24. Tris Buffered Saline containing Tween-20 (TBST, 1×); 1× solution in water with 0.1% Tween-20.

25. Blocking solution: 5% nonfat dry milk solution in 1× TBST.

26. Primary antibody diluent: 5% bovine serum albumin solution in 1× TBST.

27. CD74 antibody.

28. HRP-linked secondary antibody.

29. Enhanced chemiluminescence (ECL) substrate.

30. Kim Wipes.

31. Plastic wrap.

32. Film cassette.

33. Autoradiography film.

34. Film developer.

2.4 Immunohisto-chemistry

1. Antigen Unmasking Solution: 1% solution in water, freshly prepared (see **Note 3**).

2. Hydrogen peroxide: 3% solution in water, freshly prepared.

3. Phosphate buffered saline (PBS, 20×): 800 mL water, 160 g NaCl, 4 g KCl, 28.8 g Na_2HPO_4, 4.8 g KH_2PO_4. On a stir plate, ensure all reagents are dissolved, adjust pH to 7.4, and Qs to 1 L.

4. PBS ($1\times$): 950 mL water, 50 mL $10\times$ PBS.

5. VECTASTAIN Elite ABC HRP Kit.

6. MIF and CD74 antibodies.

7. ImmPACT DAB Peroxidase Substrate: 1 drop (~30 µL) IMMPACT DAB Chromogen concentrate in 1 mL ImmPACT DAB Diluent.

8. Hematoxylin.

9. Vectamount.

10. Cover slips.

---

3   Methods

3.1   Tissue Processing

1. The biopsy tissue is placed directly into 10% neutral buffered formalin after being obtained from the clinic.

2. Store the tissue in 10% NBF for at least 24 h to allow permeation of the tissue (*see* **Note 4**).

3. Transfer the tissue to 70% ethanol for at least 24 h (*see* **Note 4**).

4. Label a tissue embedding cassette with pencil (*see* **Note 5**).

5. Pour the tissue and ethanol over a beaker into the tissue embedding cassette. The cassette will catch the tissue and the ethanol will flow through into the beaker. Place the cassette containing the tissue into a new beaker of 70% ethanol where it can be stored at 4 °C until processing.

6. Place tissue embedding cassette into the tissue processor and start the program, sequentially changing reagents:

7. 70% ethanol at room temperature for 30 min (*see* **Note 6**).

8. 80% ethanol at room temperature for 30 min.

9. 95% ethanol at room temperature for 20 min.

10. 95% ethanol at room temperature for 20 min.

11. 95% ethanol at room temperature for 15 min.

12. 100% ethanol at room temperature for 15 min.

13. 100% ethanol at room temperature for 15 min.

14. 100% ethanol at room temperature for 15 min.

15. Xylene substitute at room temperature for 20 min.

16. Xylene substitute at room temperature for 15 min.

17. Paraplast at 57 ° C for 20 min.

18. Paraplast at 57 ° C for 20 min.

19. At Embedding station fill a mold with melted paraffin and carefully transfer the tissue to the mold.

20. Position the tissue in desired orientation (*see* **Note 7**).

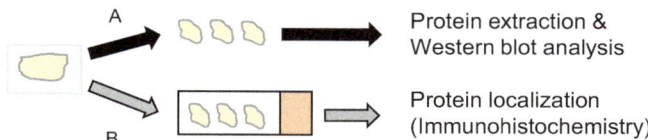

Fig. 1 Schematic representation of tissue sample processing of paraffin-embedded tissue blocks for simultaneous assessment of protein quantitation by (**a**) Western blot analysis and (**b**) protein localization using immunohistochemical techniques

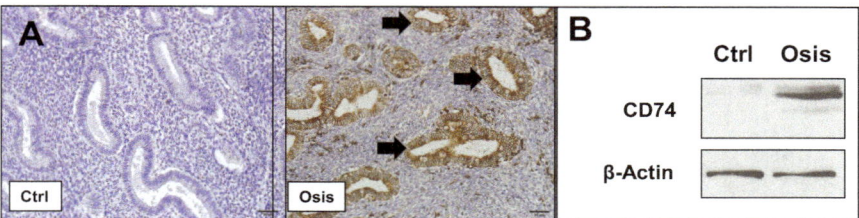

Fig. 2 Immunohistochemical localization and Western analysis of CD74 in eutopic endometrium from women without and with endometriosis. Paraffin-embedded eutopic endometrial tissue from women without (Ctrl) and with (Osis) endometriosis were subjected to immunohistochemical localization (**a**) for CD74 while additional tissue sections from paraffin-embedded blocks were subjected to protein extraction and Western blot analysis (**b**). CD74 is localized to endometrial epithelial glands (arrows) in women with endometriosis (Osis) and this pattern of expression is replicated in Western blot analysis providing both localization and quantitative data for CD74 from the same tissue specimen

21. Slide cassette over the mold and fill completely with melted paraffin.

22. Place the mold with tissue and cassette to the cooling portion of the embedding station.

23. Once the paraffin is completely solidified, remove the mold.

24. Trim excess paraffin on the outside of the mold with a razor blade.

25. Store at 4 °C.

26. This formalin-fixed paraffin-embedded tissue can now be used for protein isolation, western blot, and immunohistochemistry (Figs. 1 and 2).

3.2 Protein Extraction and Quantification

Before beginning, ensure components of the kit are made and a heat block is set to 100 °C and a shaking incubator is set to 80 °C (*see* **Note 8**).

1. Using a microtome, cut 3 serial sections from the block (*see* **Note 9**).

2. Place the sections into a 1.5 mL collection tube supplied with the kit.

3. Add 1 mL xylene to the tube, vortex vigorously for 10 s, and incubate at RT for 10 min.

4. Centrifuge for 2 min at full speed and discard the supernatant.

5. Repeat **steps 3** and **4** one additional time.

6. Add 1 mL of 100% ethanol to the pellet, mix by vortexing, and incubate at RT for 10 min.

7. Centrifuge for 2 min at full speed and discard the supernatant, being careful not to disrupt the pellet.

8. Repeat **steps 6** and **7** one additional time.

9. Add 1 mL of 96% ethanol to the pellet, mix by vortexing, and incubate at RT for 10 min.

10. Centrifuge for 2 min at full speed and discard the supernatant, being careful not to disrupt the pellet.

11. Repeat **steps 9** and **10** one additional time.

12. Add 1 mL of 70% ethanol to the pellet, mix by vortexing, and incubate at RT for 10 min.

13. Centrifuge for 2 min at full speed and discard the supernatant, being careful not to disrupt the pellet.

14. Repeat **steps 12** and **13** one additional time.

15. Prepare Extraction Buffer EXB by adding 6 μL of 2-mercaptoethanol to 94 μL of Extraction Buffer EXB Plus (*see* **Note 1**).

16. Add 100 μL of Extraction Buffer EXB to the pellet and mix by vortexing.

17. Seal the tube with a Collection Tube Sealing Clip supplied with the kit.

18. Incubate on ice for 5 min and mix again by vortexing.

19. Incubate the tube at 100 °C for 20 min.

20. Incubate the tube at 80 °C for 2 h with agitation at 750 rpm.

21. Incubate the tube at 4 °C for 1 min and remove the Collection Tube Sealing Clip.

22. Centrifuge for 15 min at $14,000 \times g$ at 4 °C.

23. Transfer the supernatant, which contains the extracted protein, to a new microcentrifuge tube and store on ice for the remainder of the protocol.

24. Set up glass tubes for 4 BSA standards and the newly extracted protein, in duplicate. Also, include one tube to use as a blank.

25. Add 780 μL of water to each tube.

26. Add 5 μL of standard to each standard tube and 5 μL protein to each protein tube. Do not add anything to the blank tube.

27. Add 200 μL of protein assay reagent to each tube.

28. Vortex each tube.

29. Measure the absorbance of each tube at 595 nm by pouring it into a plastic cuvette and placing into a spectrophotometer.

30. Average the duplicate absorbances.

31. In Excel, plot the standard absorbances in a linear line chart.

32. Determine the protein concentration by solving for X in the line equation, where Y is the protein absorbance. The resulting value for X is the protein concentration in μg/μL.

3.3 Western Blot Analysis

1. Clean each glass plate with 70% ethanol and dry with Kim Wipe.

2. Put plates into Mini-PROTEAN Tetra Cell Casting frame and secure into stand (*see* **Note 10**).

3. Prepare 12% separating gel by mixing 3 mL water, 3 mL 40% Acrylamide/Bis solution, 4 mL 1 M Tris–HCl pH 8.8, 100 μL 10% SDS, 5 μL TEMED, and 100 μL 10% APS. Swirl gently and pipet between the glass plates until 2/3 full. Add 400 μL isopropanol on top of the separating gel.

4. Allow approximately 30 min for the separating gel to solidify. Pour off isopropanol and rinse gently with water.

5. Prepare 8% stacking gel by mixing 3.5 mL water, 1 mL 40% Acrylamide/Bis solution, 400 μL 1 M Tris–HCl pH 6.8, 50 μL 10% SDS, 2.5 μL TEMED, and 50 μL 10% APS. Swirl gently and pipet on top of the solidified separating gel, filling the plates to the top. Carefully place comb between the glass plates.

6. Allow approximately 30 min for the gel to solidify.

7. Remove the plates from the casting apparatus and gently pull the comb out while running the plates under deionized water.

8. Place the plates in the electrode assembly with the short plate facing inward. Create an inner chamber by using a buffer dam on the opposite side of the gel. Place the electrode assembly into the buffer tank.

9. Fill the inner chamber completely with $1\times$ running buffer and fill the outer chamber at least 2/3 of the way full.

10. Add 30 μg protein to a new microcentrifuge tube and then add 10 μL sample buffer (*see* **Note 11**).

11. Heat at 95 °C for 5 min and then centrifuge briefly.

12. Load 10 μL of protein standard into the first well and load the entire protein/sample buffer sample into the following wells.

13. Perform electrophoresis with the power supply set at 70 V until the dye front has reached the bottom of the gel. Once the electrophoresis is complete, stop the power supply and remove the gel.

14. Pry apart the glass plates carefully using a gel releasing tool and cut off the stacking gel and discard appropriately. Carefully release the gel into a dish filled with cold $1\times$ transfer buffer.

15. Assemble a transfer cassette in a large dish filled with cold $1\times$ transfer buffer. The order for cassette components should be as follows: white end of cassette—foam pad—filter paper—nitrocellulose membrane—gel—filter paper—foam pad—black end of cassette (*see* **Note 12**).

16. Place the cassette into transfer core with the black end of the cassette facing the black part of the core and the white part of the cassette facing the red side of the core. Place an ice pack in the tank and fill with $1\times$ transfer buffer. Pack ice around the outside of the tank (*see* **Note 13**).

17. Transfer for 60 min at 100 V.

18. Remove the cassette and lay it flat with the white side down. Open the cassette and using forceps, move the membrane to a tray filled with $1\times$ TBST. Discard the gel and filter papers and rinse all gel running and transfer components.

19. Incubate in blocking solution for 30 min.

20. Prepare CD74 and/or MIF primary antibody solution in 5% BSA (*see* **Note 14**).

21. Incubate in primary antibody overnight at 4 °C.

22. Wash in $1\times$ TBST $3\times$ for 5 min each.

23. Incubate in secondary antibody at a 1:5000 ratio for 60 min.

24. Wash in $1\times$ TBST $3\times$ for 15 min each.

25. Add freshly prepared ECL detection reagent and swirl for 60 s.

26. Drain excess detection reagent by holding one side with forceps and leaning the opposite side gently on a Kim Wipe.

27. Wrap the membrane in plastic wrap and smooth out any detection reagent still apparent. Tape the plastic wrap into a film cassette.

28. In a dark room, place autoradiography film in the cassette, close, and expose for 1 min. Exposure times should be determined empirically for optimal results.

3.4 Immunohisto-chemistry

1. Cut sections 9 μm thick on microtome.

2. Float sections in tissue flotation water bath.

3. Pick up sections with charged slide and label with pencil (*see* **Note 2**).

4. Place on 37 °C slide warmer for 24 h.

5. Heat Antigen Unmasking Solution to 95 °C (*see* **Note 15**) and place a coplin jar of $1\times$ PBS on ice.

6. Label each slide with pencil with the date and the antibody.

7. Deparaffinize slides by placing them in xylenes, 2× for 5 min each (*see* **Note 16**).

8. Hydrate slides through an ethanol series of 100%, 95%, and 70% ethanol for 3 min each.

9. Wash in tap water for 5 min.

10. Incubate in Antigen Unmasking Working Solution for 10 min at 95 °C (*see* **Note 17**).

11. Wash in ice-cold 1× PBS for 5 min.

12. Quench endogenous peroxidases by submerging slides in 3% H_2O_2 for 6 min.

13. Wash in 1× PBS 3× for 5 min each.

14. Dry each slide around the tissue and draw a generous circle around the tissue with a PAP pen to create a hydrophobic barrier (*see* **Note 18**).

15. Incubate in diluted normal serum for 30 min at room temperature.

16. Incubate in CD74 primary antibody, diluted in normal serum, for 1 h at room temperature (*see* **Note 19**).

17. Wash in 1× PBS 3× for 5 min each.

18. Incubate in diluted biotinylated secondary antibody for 30 min.

19. Prepare ABC reagent, so that it sits for at least 30 min before use.

20. Wash in 1× PBS 2× for 5 min each.

21. Incubate in ABC Reagent for 30 min.

22. Wash in 1× PBS 2× for 5 min each.

23. Develop in freshly prepared ImmPACT DAB for 1–15 min (*see* **Note 20**).

24. Rinse in tap water.

25. Counterstain with Hematoxylin for 40 s.

26. Wash in tap water for 5 min.

27. Dehydrate slides through an ethanol series of 70%, 95%, and 100% ethanol for 1 min each (*see* **Note 21**).

28. Clear slides in xylenes 2× for 5 min each.

29. Allow slides to completely dry.

30. Mount with 35 μL VectaMount and cover slip.

4   Notes

1. A master mix of Extraction Buffer EXB can be made if there are multiple tubes in this same ratio. The master mix should be 6% 2-mercaptoethanol in Extraction Buffer EXB Plus.

2. Protein assay kits are available from several vendors. Follow protocol-specific instructions for performing the assay to determine protein concentrations.

3. Antigen Unmasking solutions are available from a number of companies and typically compose of tris-based or citrate-based. They are designed to reveal or unmask antigens when tissue sections are incubated in these solutions at high temperature.

4. The size of the tissue varies. Larger tissue will require more time for complete permeation. Allow 24 h at minimum and 72 h at maximum.

5. All labeling should be done with pencil. Tissue embedding cassettes and slides will be fully submerged in xylenes and ethanol causing anything written with marker to be washed away.

6. Room temperature is between 22 °C and 24 °C.

7. At this point while the paraffin is hot, the tissue can be moved around and placed wherever desired in the block and in whatever orientation desired. Our preferred method is to place the tissue toward the bottom of the mold and in the center. This provides consistency in locating the tissue once the block is solidified. If there is a certain part of the tissue that should be sectioned first, this should be placed "facing down" as the bottom of the mold will become the top of the block.

8. The Qproteome FFPE Tissue Handbook indicates a Thermomixer set to 80 °C and 750 rpm be used here. Placing the tube in an 80 °C incubator with a rotator at 750 rpm will work as well.

9. It is important to section far enough into the block to get a good cross section of tissue. If the first few sections are collected, it likely will not be representative of the whole sample. If collecting from a block that has been cut before and has been stored exposed to air, throw away the first 4 sections and start collecting after that. We found 3 sections to provide sufficient protein, but some trial and error may be required to find the optimal protein concentration desired based on the size of the tissue sample.

10. Place one spacer plate and one short plate together, creating a 1.5 mm gap between the plates and slide them into the gel casting frame. Before locking the frame, make sure the bottoms

of the plates are against a flat surface. After locking the frame, inspect the bottom to make sure the plates are flush. This will help ensure that there are no leaks with the gel is poured. Place a gray gasket on the gel casting stand and snap the plates and frame into the stand.

11. The amount of protein to load in the gel should be determined by the user. We find 30 μg is usually sufficient.

12. When placing the gel on the nitrocellulose membrane, it should be placed in the same orientation as when it was loaded. Care should be taken to smooth out any air bubbles between the membrane and the gel at this step by gently coaxing them out with a gloved finger, although generally speaking, when the cassette is closed any air bubbles should escape out the sides.

13. The transfer generates heat, but the membrane and gel need to be kept cool. This is the reason for packing with ice. The power supply and entire apparatus can also be transferred to a refrigerator or cold room and the transfer be performed there.

14. Concentration of the antibody should be empirically determined. For CD74, we use ab64772 from Abcam at a concentration of 0.5 μg/mL. For MIF, we use sc20121 from Santa Cruz Biotechnology at a concentration of 1:250.

15. Fill a 500 mL glass beaker up ¼ of the way with water and place on a hot plate. Prepare the Antigen Unmasking Solution in a coplin jar and place the coplin jar in the water on the hot plate. Place a thermometer in the coplin jar. Cover the beaker with foil. Once this reaches around 55 °C proceed with the deparaffinization and dehydration.

16. The deparaffinization and rehydration steps should be performed in a hood.

17. If the Antigen Unmasking Solution has not yet rose to 95 °C by the time this step is reached, leave the slides in water until it does. Once it reaches 95 °C, remove the foil and thermometer. Carefully remove the coplin jar (we prefer plastic as it does not feel as hot as glass) and transfer the slides. Place the coplin jar containing the slides back into the boiling water and re-cover with foil. Start the timer for 10 min.

18. This step should be done one slide at a time. Draw the pap pen circle and then add the diluted normal serum before moving on to the next slide to avoid the tissue drying out.

19. Specific dilution and duration of primary antibody incubation should be determined empirically by the user. Incubation time can be either room temperature for 1 h or 4 °C overnight. For CD74 we used ab9514 from Abcam at a concentration of 2 μg/mL.

20. This process should be closely watched under a microscope to determine the optimal time for DAB development. Once a time is determined, this should be kept consistent for each slide using the same antibody.

21. Dehydration, clearing, and drying of the slides should all be performed in the hood.

References

1. David JR (1966) Delayed hypersensitivity in vitro: its mediation by cell-free substances formed by lymphoid cell-antigen interaction. Proc Natl Acad Sci U S A 56:72–77

2. Bloom BR, Bennett B (1966) Mechanism of a reaction in vitro associated with delayed-type hypersensitivity. Science 153:80–88

3. Leng L, Metz CN, Fang Y, Xu J, Donnelly S, Baugh J, Delohery T, Chen Y, Mitchell RA, Bucala R (2003) MIF signal transduction initiated by binding to CD74. J Exp Med 197:1467–1476

4. Bozzi F, Mogavero A, Varinelli L, Belfiore A, Manenti G, Caccia C, Volpi CC, Beznoussenko GV, Milione M, Leoni V, Gloghini A, Mironov AA, Leo E, Pilotti S, Pierotti MA, Bongarzone I, Gariboldi M (2017) MIF/CD74 axis is a target for novel therapies in colon carcinomatosis. J Exp Clin Cancer Res 36:16

5. McClelland M, Zhao L, Carskadon S, Arenberg D (2009) Expression of CD74, the receptor for macrophage migration inhibitory factor, in non-small cell lung cancer. Am J Pathol 174:638–646

6. Takahashi K, Koga K, Linge HM, Zhang Y, Lin X, Metz CN, Al-Abed Y, Ojamaa K, Miller EJ (2009) Macrophage CD74 contributes to MIF-induced pulmonary inflammation. Respir Res 10:33

7. Valiño-Rivas L, Baeza-Bermejillo C, Gonzalez-Lafuente L, Sanz AB, Ortiz A, Sanchez-Niño MD (2015) CD74 in kidney disease. Front Immunol 26:48

8. Nothnick WB, Falcone T, Olson MR, Fazleabas AT, Tawfik OW, Graham A (2018) Macrophage migration inhibitory factor receptor, CD74 is overexpressed in human and baboon (Papio Anubis) endometriotic lesions and modulates endometriotic epithelial cell survival and interleukin-8 expression. Reprod Sci 11:1557–1166

Chapter 12

Methods to Determine the Effects of MIF on In Vitro Osteoclastogenesis Using Murine Bone Marrow-Derived Cells and Human Peripheral Blood Mononuclear Cells

Ran Gu

Abstract

Osteoclasts are the only cells that are capable of resorbing bones, and they are involved in multiple diseases and disorders. This chapter will describe several in vitro osteoclastogenesis methods, which allows further investigation of molecular mechanisms of osteoclastogenesis in normal physiological and disease conditions. This chapter includes a protocol for isolating osteoclast progenitors from mouse bone marrow and human peripheral blood, as well as obtaining murine osteoblasts for the coculture system. Furthermore, culture and identification of multinucleated osteoclasts in vitro is also described in this chapter.

Key words Osteoclast, Osteoclastogenesis, MIF, RANKL, M-CSF, TRAP, Actin ring, Bone resorption

1 Introduction

Osteoclasts are hematopoietic, large multinucleate cells, responsible for bone resorption, which play an important role in maintaining bone homeostasis during normal physiological conditions [1]. Alterations of osteoclast differentiation and activity have been reported in multiple diseases, including musculoskeletal diseases [2], autoimmune diseases [3], and cancers [4]. Culturing osteoclasts in vitro allows for the investigation of the effects of drugs and/or small molecule inhibitors on osteoclastogenesis, further allowing the interrogation of underlying molecular mechanisms. Macrophage migration inhibitory factor (MIF) is a unique pro-inflammatory molecule that is induced by glucocorticoids (GC), and plays an indispensable role in the pathogenesis of rheumatoid arthritis (RA) [5]. The bone erosion in RA is the result of the activity of osteoclasts, and MIF inhibition has been reported to reduce synovitis and bone erosion in multiple animal models of RA [6, 7]. The ablation of MIF results in decreased arthritis severity,

James Harris and Eric F. Morand (eds.), *Macrophage Migration Inhibitory Factor: Methods and Protocols*,
Methods in Molecular Biology, vol. 2080, https://doi.org/10.1007/978-1-4939-9936-1_12,
© Springer Science+Business Media, LLC, part of Springer Nature 2020

and less bone erosion by regulating osteoclastogenesis [8]. Furthermore, the in vitro osteoclastogenesis method has been utilized to demonstrate that MIF blockade also suppresses receptor activator of nuclear factor kappa-κ ligand (RANKL)-induced NF-κB and NFATc1 activity during osteoclastogenesis [8]. Several methods of obtaining osteoclast progenitors are available for the in vitro osteoclast differentiation cultures.

Previously, in vitro osteoclastogenesis was induced with the support of osteoblasts, until the discovery of RANKL, which allows for a pure culture system [9]. Macrophage colony-stimulating factor (M-CSF) is critical for maintaining the homoeotic lineage proliferative ability, and RANKL is the inductive stimuli of osteoclast differentiation, both of which are indispensible during osteoclastogenesis in vitro [10, 11]. To date, commonly used osteoclast progenitors for in vitro osteoclastogenesis are murine bone marrow cells, RAW264.7 cells, and human peripheral blood mononuclear cells (PBMCs) [8, 12]. At the end of culture, multinucleated (more than 3 nuclei) type-5 tartrate-resistant acid phosphatase (TRAP) staining-positive cells are commonly accepted as osteoclasts. Bone resorption ability is a unique feature of osteoclasts. To achieve this, osteoclasts alter their cytoskeleton structures upon attachment to the bone surface, and form a resorption area, termed the sealing zone [13]. The alteration of osteoclast actin structure, named the actin ring, can be observed in vitro via fluorescence staining [14]. Furthermore, the in vitro bone resorption assay is a useful method to identify osteoclast proliferation, differentiation, and activation. However, it is time-consuming and laborious [15].

This chapter will describe procedures for obtaining osteoclast progenitors from murine bone marrow, a macrophage cell line and human peripheral blood. Several common in vitro osteoclastogenesis methods will be included in this chapter. They are the osteoblast-osteoclast progenitor coculture assay; in vitro osteoclastogenesis assays using bone marrow cells or RAW 264 cells (murine macrophage cell line) as progenitors, with the stimulation of RANKL, and a bone explant assay. In addition, assays for the detection of osteoclasts will also be described in this chapter. This includes the TRAP histochemical staining, actin ring fluorescence labeling, and a quick method for determining osteoclast resorption pits.

2 Materials

2.1 Animals and Cells

1. RAW 264.7 macrophage cell line can be purchased from ATCC.

2. C57BL/6J mice, 6–8 weeks old (*see* **Note 1**).

3. Neonatal mouse pups (day 1–2) from the C57BL/6J stain for primary osteoblast extraction.

2.2   Instrument

All instruments should be sterilized prior to use.

1. Cell Strainers, 70 μm, sterile.

2. Dentine slides, sterilized by autoclave or soaking in 70% ethanol for at least 30 min and air-dried.

3. 50 mL centrifuge tubes.

4. Glass coverslips, sterilized by autoclaving or soaking in 70% ethanol for at least 30 min and air-dried.

5. Paper wipes (e.g., KimWipes).

6. Multi-well tissue culture plates: 96-well plates, 48-well plates, and 24-well plates.

7. Sterile 10-cm petri dishes.

8. Surgical equipment: scalpels, scissors, and forceps.

9. 10 mL syringes with 27 gauge (27 g) needles for bone marrow flush.

2.3   Media and Solutions

1. Sterile phosphate-buffered saline (PBS), pH 7.4, sterile.

2. 50 mg/mL Ascorbic acid dissolved in sterile PBS (*see* **Note 2**).

3. Complete culture medium (cCM): Minimal essential medium-alpha (α-MEM) supplemented with 10% fetal bovine serum (FBS), 50 U/mL penicillin, 50 μg/mL streptomycin, and 2 mM L-glutamine.

4. Digestion solution for calvarial osteoblasts: dissolve 1 mg of collagenase type II and 2 mg of Dispase per 1 mL of PBS. Make fresh and filter sterilize, and keep on ice prior to use.

5. 70% Ethanol.

6. Hanks' Balanced Salt Solution (HBSS), sterile.

7. Human recombinant RANKL: 1 mg/mL stock solution in distilled H_2O.

8. Mouse recombinant M-CSF: 100 μg/mL stock solution in sterile PBS.

9. 14 μM stock solution of fluorescently labeled phalloidin (e.g., rhodamine-phalloidin) reconstituted in methanol

10. TRAP staining fixation solution: 4% paraformaldehyde by diluting 37% formaldehyde stock solution in PBS.

11. TRAP permeabilization wash buffer: 1:1 (volume/volume) acetone and methanol.

12. TRAP histochemical stain buffer: 50 mM sodium acetate, 40 nM potassium sodium tartrate in distilled H_2O, adjusted

to pH 5 with glacial acetic acid. This solution can be stored at 4 °C for several months.

13. TRAP staining solution: 5 mg Naphthol AS-MX phosphate dissolved in 0.5 mL of N,N-dimethylformamide (DMF), followed by addition of 50 mL of TRAP histochemical stain buffer from 8. 15 mg of Fast Red Violet LB Salt is then added to the solution and mixed well. The solution can be stored at 4 °C for several weeks.

14. Trypsin solution: 0.25% trypsin; 0.1% EDTA in PBS, sterile.

3   Methods

3.1   Isolating Osteoclast Precursors and Primary Osteoblasts

3.1.1   Isolation of Mouse Bone Marrow Cells

1. Euthanize the mouse with CO_2 and spray the abdominal region and hind limbs with 70% ethanol.

2. Make a scission at the lower abdominal region, near the hind limbs. Remove the skin and muscles from both hind limbs. Dislocate the limb by cutting at the pelvic-hip joint and the ankle joint. Do not separate the femur and tibia, to avoid potential contamination.

3. Trim off large muscles with scissors, followed by further cleaning with paper wipes sprayed with 70% ethanol for further cleaning of the remaining muscles around the bones.

4. Transfer the clean limb bones to a 50 mL centrifuge tube containing pre-chilled sterile PBS, and place the tube on ice.

5. Bring the tube to a sterile tissue culture hood, and transfer the bones to a 10 cm-petri dish containing cold HBSS.

6. Separate femur and tibia at the knee joint.

7. Once the femur and tibia are separated, cut at both ends of each bone.

8. Flush bone marrow out with 10 mL cold HBSS using a 10 mL syringe with the 27 g needle into a sterile 10 cm petri dish. If bone marrow is not completely removed (bone is not white and clear looking), flush again with another 5 mL of cold HBSS.

9. Agitate cells by pipetting several times.

10. Place a cell strainer on top of a new 50 mL centrifuge tube, and let the HBSS solution containing bone marrow cells pass through.

11. Centrifuge the tube from **step 10** at 400 × g for 5 min, the bone marrow cells are pelleted on the bottom and ready for use.

1. Euthanize the neonatal mice by decapitation, and sterilize the skin with 70% ethanol.

2. Use sterile scissors and forceps to remove the calvaria by first removing the scalp, followed by cutting away the calvaria from the base of the skull. Any extraneous fibrous or brain tissue should be carefully trimmed away. Transfer the calvaria to a 50 mL Falcon tube containing cold sterile PBS, and place the tube on ice until ready for the next step. This procedure is repeated for the remaining mice.

3. After collecting all the calvariae, bring the tube to a clean tissue culture hood. Discard the PBS, and replace with 10 mL of the digestion solution. Incubate the tube at 37 °C with agitation for 5 min, and discard the supernatant by pipetting.

4. Add another 10 mL of fresh digestion solution to the tube, and incubate at 37 °C for 20–30 min with agitation (*see* **Note 3**).

5. Collect the supernatant containing released calvarial cells to a new 50 mL centrifuge tube containing 5–10 mL cCM to inactivate the proteolytic enzymes. The tube should be kept on ice until ready to use.

6. Repeat **steps 4** and **5** again.

7. Centrifuge the tube at $400 \times g$ for 5 min after collecting all the cell-containing supernatant. Discard the supernatant after centrifuge, and the primary osteoblasts are pelleted on the tube bottom.

8. Resuspend the primary osteoblast pellet with cCM, and count cells.

9. Seed cells at approximately 5×10^6 cells in 10 mL of cCM in a 10 cm petri dishes, and incubate at 37 °C in a humidified incubator with 5% CO_2 for 2–3 days, to expand the osteoblast population for further assays.

3.2   Culturing Osteoclasts In Vitro

3.2.1   Osteoclastogenesis with Coculture of Bone Marrow Cells and Osteoblasts

1. Harvest the previously prepared primary osteoblasts with trypsin solution, and seed the osteoblasts accordingly (2×10^4 cell/well) with 0.2 mL cCM media in a 48-well plate (*see* **Note 4**).

2. Let osteoblasts settle down by incubating at 37 °C in a humidified incubator with 5% CO_2 for 2–3 h, prior to the addition of osteoclast progenitor cells.

3. Add osteoclast progenitor cells (murine bone marrow cells or RAW 264.7 cells) accordingly (1×10^5/well). The total medium volume should be 0.4 mL/well (*see* **Note 5**).

4. Add 50 µg/mL ascorbic acid and 10^{-8}M dexamethasone to the culture. Incubate the cocultured cells for 7 days, with complete medium change on day 3.

5. At the end of the culture period, the cells are subjected to TRAP histochemical staining, as discussed in Subheading 3.3.

3.2.2 Osteoclastogenesis with Bone Marrow Derived Precursors

1. Resuspend cells obtained from **step** 11 in Subheading 3.1.1 with cCM and perform cell count.

2. Seed cells into a 96-well plate with 0.1 mL cCM at 1×10^4 cells/well.

3. Add osteoclastogenic stimulus by diluting RANKL and M-CSF in 0.1 mL cCM/well. The final volume of each well should be 0.2 mL. The final culturing concentration of M-CSF is 30 ng/mL and the final concentration of RANKL is 50 ng/mL (*see* **Note 6**).

4. Perform a complete medium change on day 3.

5. On day 7, the cells are subjected to TRAP histochemical staining.

3.2.3 Osteoclastogenesis with Bone Explantation Method

1. The leftover bone shaves from **step 8**, Subheading 3.1.1 can be used for the bone explanation method. After flushing bone marrow, cut the bone shaves into four pieces, making each bone fragment similar in size.

2. Flush the bone fragments with HBSS using a 10 mL syringe with a 27 g needle several times, making sure all the bone marrow is removed.

3. Transfer bone fragments into a 12-well plate, making sure that each well has similar amount of bones.

4. Culture the bone shaves with cCM supplemented with 100 mM ascorbic acid and 1 ng/mL M-CSF (final concentrations) for 7 days, with complete medium change on days 2, 4, and 6 (*see* **Note 7**).

5. On day 7, remove the bone fragments. Add 50 ng/mL RANKL (final concentration) to the culture medium to induce osteoclast formation.

6. Incubate this culture at 37 °C in a humidified incubator with 5% CO_2 for another 7 days, with a complete medium change on day 10.

7. On day 14, terminate experiment and cells are subjected to TRAP histochemical staining.

3.2.4 Using RAW 264.7 Cell-Lines as Osteoclast Progenitors

1. On day 0, seed RAW 267.4 cells in 96-well plates at 5×10^4 cells/well, with cCM at 37 °C in a humidified incubator with 5% CO_2 overnight.

2. The next day, add 100 ng/mL RANKL (final concentration) to the culture, and incubate the cells for 7 days, with a complete medium change on day 3 (*see* **Note 8**).

3. Experiment is terminated on day 7 and cells are subjected to TRAP histochemical staining.

3.2.5 Human Peripheral Blood Mononuclear Cells (PBMCs) as Osteoclast Progenitors

This method is modified from Eeles et al. [12] and Wu et al. [16].

1. Dilute human peripheral blood with sterile PBS at 1:1 ratio.

2. Slowly layer the diluted blood to the pre-warmed Histopaque-1077; make sure there no blood mixed with the Histopaque.

3. Centrifuge at $1000 \times g$ for 20 min at 18 °C.

4. Collect the buffy layer of white blood cells, and transfer it to a new 50 mL centrifuge tube.

5. Resuspend the cell pellet in sterile PBS and centrifuge again at $650 \times g$ for 5 min.

6. Discard supernatant and resuspend cells with cCM.

7. Seed cells at 1×10^6/well in 96-well plate in 0.1 mL volume, and add another 0.1 mL of cCM supplemented with 30 ng/mL M-CSF (final concentration) and 100 ng/mL RANKL (final concentration) for 14 days, with complete medium change every 3 days.

8. At the end of each experiment, fix cells and stain for TRAP.

3.3 Identification of Osteoclasts

3.3.1 Bone Resorption Assay

1. Mark the bottom of each dentine slice with pencil.

2. Prepare dentine slices by soaking in 70% ethanol for at least 30 min, followed by 2 washes with sterile PBS. The slices are dried in a tissue culture hood until they are ready for use.

3. Place the dentine slices into a 96-well plate, one slice per well, with the marked side face down.

4. Osteoclasts are generated as described in Subheading 3.2.2, except the experiment is 14 days instead of 7 days, with complete medium change every 3 days.

5. At the end of experiment, remove cells from the dentine slices by treating the slices with chloroform:methanol (1:3, v/v) under agitation for 2–3 min, followed by a wash of 100% methanol under agitation for 30–60 s.

6. Discard methanol and transfer dentine slices into a glass beaker and let them air-dry.

7. For quantification of resorption, the dentine slices can be stained with permanent marker, and wiped to remove marker from the surface of the bone/dentine. The stain remains in the pits rendering them visible. Pit area is measured under a light microscope by eye using a point-counting method.

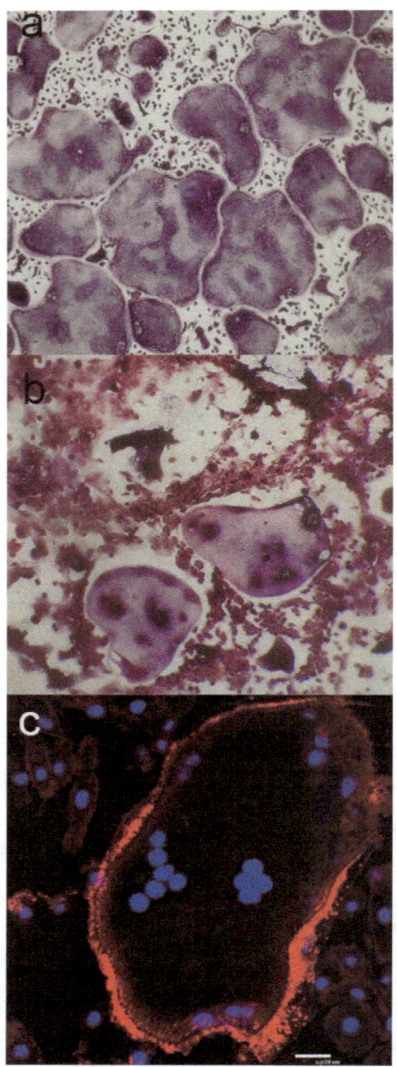

Fig. 1 Osteoclastogenesis in vitro. (**a**) Osteoclasts were induced with M-CSF and RANKL using murine bone marrow cells as progenitors. TRAP staining was performed using the Sigma TRAP staining kit. (**b**) Osteoclasts derived from the bone explant culture; osteoclasts were stained for TRAP with homemade TRAP staining solution. (**c**) Osteoclasts were induced from murine bone marrow cells and stained with rhodamine-phalloidin for actin ring formation

*3.3.2   TRAP
Histochemical Staining*

1. Remove culture medium, and rinse the cells with cold PBS for 5 min.

2. Fix cells with 4% paraformaldehyde in PBS for 5 min at RT.

3. Remove the fixation solution, and incubate the cells with cold TRAP permeabilization solution for 30 s.

4. Remove the permeabilization solution, and let plate air-dry.

5. Add TRAP staining solution to each well and incubate at RT for 10–30 min, until the appearance of pink/red color of TRAP$^+$ MNCs can be seen (Fig. 1). Check the development of color frequently.

6. Once desirable color is achieved, rinse the plate tap water, and store dry until ready for pictures.

This histochemical staining remains easily visible for at least 1 year. Alternatively, the TRAP staining kit from commercial companies also reveal excellent TRAP staining results, often staining the osteoclasts with a purple color instead of pink. Representative pictures of TRAP-positive multinucleated osteoclasts are shown in Fig. 1.

3.3.3 Actin Ring Determination with Rhodamine-Phalloidin

1. Bone marrow cells should be seeded on a glass coverslip (typically in 24-well plates and the staining process can be performed in the wells) for imaging the actin ring formation.

2. Stimulate for osteoclastogenesis as described in Subheading 3.2.2.

3. At end of experiment, remove culture medium. Wash cells with cold PBS for 5 min, and repeat this step twice.

4. Fix cells with 4% paraformaldehyde in PBS for 30 min at room temperature.

5. Permeabilize cells with 0.1% Triton-100 in PBS for 5 min at room temperature.

6. Wash with PBS for 5 min, and repeat this step twice.

7. Stain cells with 100 nM fluorescently labeled phalloidin (final concentration) for 30 min at room temperature in the dark.

8. Remove the staining solution, and wash the cells with PBS three times.

9. Put a droplet of mounting medium with DAPI on a glass slide. Remove coverslips from the plate carefully with forceps, and place the coverslip (cell side down) on the droplet of mounting medium.

10. Dry slides in dark for 30 min and seal the coverslips with nail polish, if necessary.

11. Osteoclast peripheral actin rings can be observed by confocal microscopy. A representative picture of osteoclast actin ring formation is shown in Fig. 1c.

4   Notes

1. Mice at the age of 6–8 weeks give the best osteoclastogenesis. Mice should be matched for age and gender when comparing

the osteoclastogenesis ability between WT mice and MIF knockout (KO) mice (or any other groups).

2. Ascorbic acid should be aliquoted and stored in −20 °C. The thawed aliquot should not be reused.

3. A 37 °C water bath with shaking ability is preferred.

4. The Kusa O multipotential bone marrow stromal cell line can be used as an alternative source for primary osteoblasts [17].

5. In the coculture system, WT and $Mif^{-/-}$ osteoblasts can be cultured with WT or $Mif^{-/-}$ BMs to determine (1) the effects of MIF on the supportive effects of osteoblasts for osteoclastogenesis and (2) the effects of MIF on osteoclast progenitors.

6. RANKL-induced osteoclastogenesis using bone marrow cells as osteoclast progenitors is an easy and flexible method to determine the effects of MIF on osteoclastogenesis, which can be achieved by using MIF knockout bone marrow cells, or blocking MIF with an antibody or a commercially available small molecule inhibitor, prior to the addition of the RANKL stimuli [18]. Alternatively, to examine the rescue effects of MIF, recombinant MIF protein can be added to the $Mif^{-/-}$ culture prior to RANKL.

7. The bone explantation method was established to investigate another source of osteoclast progenitors and showed excellent osteoclastogenesis. However, this is not a quantitative assay as it is hard to control the starting number of osteoclast progenitors in the bone. More information can be found from Chang et al. [19]. We have used the bone fragments from the $Mif^{-/-}$ mice and found a trend toward reduction of osteoclast formation when MIF is ablated [20].

8. MIF-blocking antibodies or small molecule inhibitors should be added 1–2 h prior to RANKL stimulation. In addition, luciferase reporter cell lines have been made in the RAW 264.7 cell lines, including the P3K cells (NF-κB reporter) [20] and NFAT-RAW reporter cell lines [8]. Addition of MIF blockade before RANKL stimulation would allow the examination of the effects of MIF on RANKL-induced molecular mechanisms during osteoclast differentiation.

References

1. Teitelbaum SL (2000) Bone resorption by osteoclasts. Science 289(5484):1504–1508

2. Manolagas SC (2000) Birth and death of bone cells: basic regulatory mechanisms and implications for the pathogenesis and treatment of osteoporosis. Endocr Rev 21(2):115–137

3. Goldring SR, Gravallese EM (2000) Pathogenesis of bone erosions in rheumatoid arthritis. Curr Opin Rheumatol 12(3):195–199

4. Olechnowicz SW, Edwards CM (2014) Contributions of the host microenvironment to cancer-induced bone disease. Cancer Res 74 (6):1625–1631

5. Morand EF, Leech M, Bernhagen J (2006) MIF: a new cytokine link between rheumatoid arthritis and atherosclerosis. Nat Rev Drug Discov 5(5):399–410

6. Santos LL et al (2008) Reduced arthritis in MIF deficient mice is associated with reduced T cell activation: down-regulation of ERK MAP kinase phosphorylation. Clin Exp Immunol 152(2):372–380

7. Herrero LJ et al (2011) Critical role for macrophage migration inhibitory factor (MIF) in Ross River virus-induced arthritis and myositis. Proc Natl Acad Sci U S A 108 (29):12048–12053

8. Gu R et al (2015) Macrophage migration inhibitory factor is essential for osteoclastogenic mechanisms in vitro and in vivo mouse model of arthritis. Cytokine 72(2):135–145

9. Suda T et al (1999) Modulation of osteoclast differentiation and function by the new members of the tumor necrosis factor receptor and ligand families. Endocr Rev 20(3):345–357

10. Arai F et al (1999) Commitment and differentiation of osteoclast precursor cells by the sequential expression of c-Fms and receptor activator of nuclear factor kappaB (RANK) receptors. J Exp Med 190(12):1741–1754

11. Kim JH, Kim N (2016) Signaling pathways in osteoclast differentiation. Chonnam Med J 52 (1):12–17

12. Eeles DG et al (2015) Osteoclast formation elicited by interleukin-33 stimulation is dependent upon the type of osteoclast progenitor. Mol Cell Endocrinol 399:259–266

13. Lakkakorpi PT, Vaananen HK (1996) Cytoskeletal changes in osteoclasts during the resorption cycle. Microsc Res Tech 33 (2):171–181

14. Vaananen HK et al (2000) The cell biology of osteoclast function. J Cell Sci 113 (Pt 3):377–381

15. Bradley EW, Oursler MJ (2008) Osteoclast culture and resorption assays. Methods Mol Biol 455:19–35

16. Wu DJ et al (2014) A novel in vivo gene transfer technique and in vitro cell based assays for the study of bone loss in musculoskeletal disorders. J Vis Exp 88. https://doi.org/10.3791/51810

17. Allan EH et al (2003) Differentiation potential of a mouse bone marrow stromal cell line. J Cell Biochem 90(1):158–169

18. Al-Abed Y, VanPatten S (2011) MIF as a disease target: ISO-1 as a proof-of-concept therapeutic. Future Med Chem 3(1):45–63

19. Chang MK et al (2008) Osteal tissue macrophages are intercalated throughout human and mouse bone lining tissues and regulate osteoblast function in vitro and in vivo. J Immunol 181(2):1232–1244

20. Singh PP et al (2012) Membrane-bound receptor activator of NFkappaB ligand (RANKL) activity displayed by osteoblasts is differentially regulated by osteolytic factors. Biochem Biophys Res Commun 422(1):48–53

Chapter 13

Inducing and Inhibiting Autophagy to Investigate Its Interactions with MIF

Nadia S. Deen, Jacinta P. Lee, and James Harris

Abstract

MIF has been described to be associated with autophagy in a number of studies, but the full nature of their association is not yet clear. However, the unprecedented interest in autophagy in recent times has generated a number of tools and techniques for its study. Here, we present protocols for studying the interactions between MIF and autophagy, including for the induction and inhibition of autophagy and measuring autophagosome biogenesis and maturation.

Key words Autophagosome, LC3, Macroautophagy, Rapamycin, siRNA, SQSTM1/p62

1 Introduction

Autophagy is a conserved eukaryotic process that maintains cellular homeostasis through lysosomal degradation of long-lived proteins and damaged organelles [1]. Three forms of autophagy have been described, chaperone-mediated autophagy, microautophagy, and macroautophagy, each of which involves the delivery of cytoplasmic cargo to lysosomes for degradation. In chaperone-mediated autophagy, soluble proteins are selectively targeted via a common pentapeptide motif (KFERQ) by the chaperone protein hsc70 and directed to lysosomes [2]. Microautophagy involves the direct engulfment of cytoplasmic proteins by lysosomes [3], while macroautophagy involves the formation of a specialized double-membraned autophagosome around a portion of cytoplasm, which can include organelles, such as mitochondria [4]. This chapter presents methods for the study and manipulation of macroautophagy (hereafter referred to as autophagy).

Autophagy is regulated by multiple environmental, physiological, and biochemical sources, including amino acid starvation, growth factor withdrawal, endoplasmic reticulum stress, and mitochondrial damage. A key regulator of autophagy is the

James Harris and Eric F. Morand (eds.), *Macrophage Migration Inhibitory Factor: Methods and Protocols*,
Methods in Molecular Biology, vol. 2080, https://doi.org/10.1007/978-1-4939-9936-1_13,
© Springer Science+Business Media, LLC, part of Springer Nature 2020

mammalian/mechanistic target of rapamycin (mTOR), which, when activated, inhibits autophagy. Drugs which inhibit mTOR, such as rapamycin (sirolimus), are potent and commonly used autophagy inducers. Autophagosome formation is dependent on the formation of a complex that includes the type III PI3 Kinase (PI3K) vascular sorting protein 34 (vps34). Thus, inhibitors of PI3Ks, such as wortmannin and LY294002, as well as the more specific type III PI3K inhibitor 3-methyladenine (3-MA) can be used to inhibit autophagy [5]. In addition, autophagy is regulated by multiple cytokines; stimulated by IFNγ, TNF, IL-1, IL-2, IL-6, and TGF-β, and inhibited by IL-4, IL-10, and IL-13 [6]. Moreover, autophagy, directly or indirectly, regulates the production and secretion of numerous cytokines, including IL-1 family cytokines, IL-6, IL-17, IL-22, IL-23, IFNγ, and type I IFNs [6, 7]. Accumulating evidence suggests that MIF can modulate autoophagy. While some studies suggest that MIF induces autophagy [8–12], others have shown that autophagy is inhibited by MIF [13–15]. Thus, further studies are required to unravel the apparently complex interactions between MIF and autophagy. In addition, loss of autophagy has been shown to increase MIF release by human and mouse macrophages in response to inflammatory stimuli [16].

The most commonly used protein marker for detecting autophagy is microtubule-associated protein 1A/1B light chain 3 (MAP1LC3, hereafter referred to as LC3) [5]. The cytosolic form of LC3, known as LC3-I, conjugates with phosphatidylethanolamine (PE) to form LC3-II, which is recruited to the membrane of autophagosomes [17]. In general, it is accepted that the level of LC3II (relative to a housekeeping protein, such as β-actin or GAPDH) is proportional to the extent of autophagosome formation, and therefore the conversion from LC3-I to LC3-II detected by Western blot is used as a common readout for autophagy [5]. Autophagy is also widely measured by the expression of fluorescently labeled LC3 fusion proteins, which after being recruited to the autophagosomal membrane, can be visualized as punctate structures by immunofluorescence microscopy [5, 18] (Fig. 1). Another less commonly used, but confirmatory test for autophagy is transmission electron microscopy, by which the double-membraned autophagosomes can be detected both qualitatively and quantitatively [5]. Furthermore, the extent of autophagy is often measured by the level of common autophagy substrates, e.g., sequestosome 1 (SQSTM1/p62), an adaptor that links cellular cargos to LC3II. Since SQSTM1 is degraded in autophagosomes, the level of this protein is considered to be inversely proportional to autophagic activity [5].

Alongside autophagosome biogenesis, it is important to also take into account the fusion of autophagosomes to lysosomes and subsequent degradation of luminal contents (referred to as "autophagic flux"). This is typically done by inhibiting lysosomal

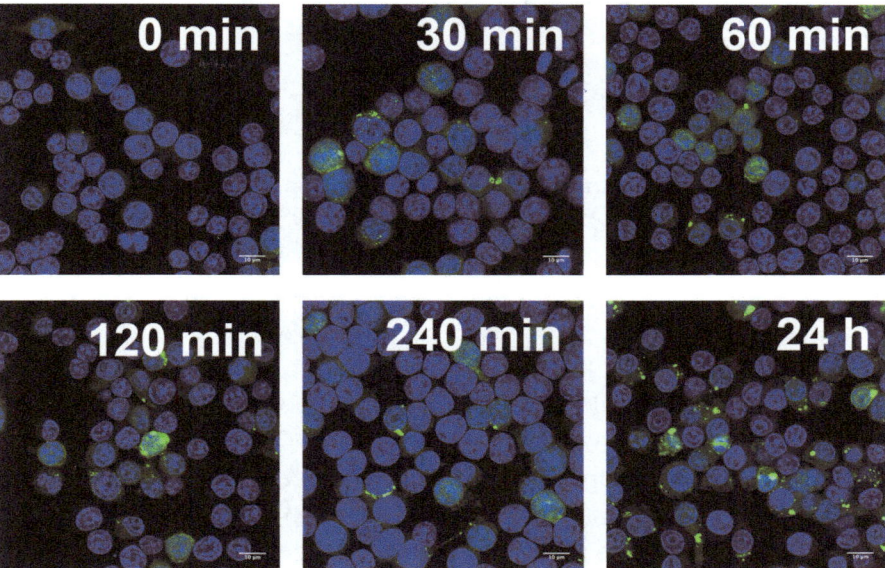

Fig. 1 LPS-induced autophagy in murine macrophages. Immortalized bone marrow-derived macrophages stably transfected with GFP-LC3 were treated with LPS (100 ng/mL) for the indicated times. GFP-LC3⁺ puncta are indicative of autophagosome formation

degradation or the fusion between autophagosomes with lysosomes, either with protease inhibitors (e.g., E64d, leupeptin), pH-modulating compounds (e.g., ammonium chloride) or lysosome fusion-blocking drugs (e.g., chloroquine, bafilomycin A1) and measuring effects on LC3 II levels by Western blot [5]. Because LC3 on the luminal side of the autophagosome is degraded, while that on the cytosolic side is recycled into the cytosol, levels of LC3 II are dynamic; increasing with increased autophagosome formation, but then decreasing with increased autophagic flux. Thus, by blocking this process, total increases in LC3 II can be more clearly seen (compared to samples not treated with flux inhibitors). Importantly, this technique allows the user to distinguish between treatments that inhibit autophagosome biogenesis from those that inhibit autophagic flux; if LC3 II levels are unaffected by flux inhibitors, this would suggest that flux is already inhibited.

In this chapter, we describe some common techniques used to regulate autophagy, including the use of autophagy-stimulating compounds and siRNA to target autophagy-specific genes. These protocols are written based on our experiences studying autophagy in macrophages, but can be applied to other cells types with appropriate optimization. This chapter does not detail assays for measuring autophagy; excellent references for this can be found elsewhere [5, 18, 19].

2 Materials

2.1 Inhibiting Autophagic Flux

1. MilliQ H_2O.

Inhibitors of autophagic flux:
2. Bafilomycin A1; stock solution of 100 μM in sterile dimethyl sulfoxide (DMSO) (*see* **Note 1**).
3. Chloroquine; stock solution of 100 μM in sterile MilliQ H_2O.
4. E64d (*see* **Note 2**); stock solution of 10 mM in sterile DMSO.
5. Leupeptin; stock solution of 10 mM in sterile MilliQ H_2O.

2.2 Inducing Autophagy in Macrophages

1. Sterile Earle's balanced salt solution (EBSS) or Hank's balanced salt solution (HBSS) (*see* **Note 3**).
2. Lipopolysaccharide (LPS) (*see* **Note 4**); stock solution of 1 mg/mL in sterile MilliQ H_2O.

mTOR inhibitors:
3. Rapamycin; stock of 10 mM in sterile DMSO.
4. Torin-1; stock of 3 mM in sterile DMSO.
5. AZD8055; 10 mM in sterile DMSO.

2.3 Inhibiting Autophagy with PI3K Inhibitors

1. Phosphate-buffered saline (PBS); 137 mM NaCl, 10 mM Na_2HPO_4, 1.8 mM KH_2PO_4, 2.7 mM KCl in milliQ water, adjust to pH 7.4 with HCl. Sterilize by autoclaving or filtration.
2. 3-Methyladenine (3-MA); stock solution of 100 mM in sterile PBS. This needs to be made immediately prior to use by heating at 50–60 °C until fully dissolved (*see* **Notes 5** and **6**).
3. Wortmannin; stock solution of 10–50 mg/mL in DMSO. This is stable for 2 months at −20 °C.
4. LY294002; stock solution of 10 mM in DMSO.

2.4 Inhibiting Autophagy with siRNA

1. RNase-free H_2O or proprietary siRNA buffer (usually diluted with RNase-free water).
2. siRNA (*see* **Note 7**); resuspend in RNase-free water of siRNA buffer to a stock concentration of 20 μM. Store at −80 °C.

2.4.1 Transfection of siRNA with a Transfection Reagent

1. Lipofectamine 2000 (*see* **Note 8**).
2. Opti-MEM cell culture medium; Opti-MEM, reduced serum medium (without serum).

2.4.2 Nucleofection of siRNA	1. Electroporation/nucleofection buffer (*see* **Note 9**).
	2. Nucleofection cuvettes (*see* **Note 10**).
	3. Complete culture medium (dependent on cell type of choice).

3   Methods

We have used the methods described below in a number of different monocyte/macrophage cell lines (e.g., THP-1, U937, RAW 264.7, J774, immortalized bone marrow-derived macrophages (iBMM)), as well as in primary human peripheral blood monocyte-derived macrophages and mouse bone marrow-derived macrophages (BMM, *see* Chapter 4 for isolation/culture protocol). Grow cells in appropriate culture dishes/plates, depending on the output being measured, e.g., flat-bottomed 96-well tissue culture plate for measurement of secreted factors by ELISA/multiplex, 12-well plates with coverslips for immunofluorescence/confocal, and 12-well or 6-well plates for Western blot.

3.1   Inhibiting Autophagic Flux

For immunoblotting and confocal analysis of conversion of LC3 I to LC3 II, it is recommended that cells are treated with and without inhibitors of autophagosome maturation (autophagic flux). This allows the experimenter to determine whether autophagosome biogenesis and/or autophagic flux is being manipulated. Alternatively, this can be used to determine effects of autophagic flux disruption on a specific output (e.g., cytokine release).

1. Treat cells with bafilomycin A1 (final concentration 100–500 nM), chloroquine (final concentration 1–10 μM), or a combination of E64d (10 μg/mL) and leupeptin (10 μg/mL) (*see* Table 2 and **Note 13**).

2. Incubate for 15–30 min at 37 °C/5% CO_2.

3. Treat with desired compounds (*see* Subheadings 3.2.1–3.3).

3.2   Inducing Autophagy in Macrophages

3.2.1 Starvation-Induced Autophagy

1. If using inhibitors (e.g., MIF inhibitors, inhibitors of autophagic flux), we recommend adding these 15–30 min prior to commencing starvation, to allow priming/uptake (*see* Subheading 3.1).

2. Remove the supernatant from cells and wash once with EBSS or HBSS.

3. Incubate in EBSS or HBSS for 0.5–6 h at 37 °C/5% CO_2. Where appropriate, add inhibitors back to cells, diluted in EBSS/HBSS.

Table 1
Autophagy-regulating compounds

	Compound	Mechanism of action	Working concentration
Autophagy inducers	Rapamycin (sirolimus)	Forms a complex with 12-kDa FK506-binding protein (FKBP12), which inhibits mTOR complex 1 (mTORC1)	0.5–10 μM
	Torin-1	An ATP-competitive inhibitor of mTORC1 and mTORC2; blocks phosphorylation of both	250–1000 nM
	AZD8055	An ATP-competitive inhibitor of the serine/threonine kinase activity of mTORC1 and mTORC2	1–1000 nM
	LPS	Multiple TNF receptor-associated factor 6 (TRAF6)-dependent K63 ubiquitination of Beclin 1, induction of ROS, mitochondrial damage and pro-inflammatory cytokines, such as TNF [16, 20, 21] (*see* **Note 11**)	1–100 nM
	Amino acid starvation	Multiple mitochondrial dysfunction, ROS production, AMP-activated protein kinase (AMPK)-dependent inhibition of mTORC1	N/A
Inhibitors of autophagosome biogenesis	3-Methyladenine (3-MA)	Preferentially inhibits type III PI3K (vps34) involved in autophagosome formation. Also inhibits other PI3K	1–10 mM
	Wortmannin	Inhibits all PI3K, including vps 34 (*see* **Note 12**)	0.1–10 μM
	LY294002	Inhibits all PI3K, including vps 34 (*see* **Note 12**)	1–100 μM
Inhibitors of autophagic flux	Bafilomycin A1	(V)-ATPase inhibitor; inhibits fusion of autophagosomes with lysosomes	100–500 nM
	Chloroquine	Inhibits autophagosome maturation by raising lysosomal/vacuolar pH	10–100 μM
	Leupeptin/E64d	Inhibitors of lysosomal proteases	10 μg/mL each

3.2.2 Inducing Autophagy with mTOR Inhibitors

1. Where appropriate, add inhibitors to cells 15–30 min prior to addition of mTOR inhibitor (*see* Subheading 3.1).

2. Add rapamycin at a final concentration of 0.5–10 μM or Torin 1 at a concentration of 250–1000 nM or AZD8055 at a concentration of 1–1000 nM (*see* **Note 13** and Table 1).

3. Incubate for 1–24 h at 37 °C/5% CO_2.

Table 2
Methods for verification of siRNA knockdown

Functional readout	Technique	Protocol
Knockdown of target gene	PCR	1. Following transfection, seed cells in 96-well plates (100 µL per well, 1–3 wells per transfection) and incubate for 24–96 h at 37°C/5% CO_2 2. Lyse cells and extract DNA/convert to RNA 3. Run PCR/qPCR for target gene
Knockdown of gene product (protein)	Western blot	1. Following transfection, seed cells in 12-well plates (1 mL per well, 1 well per transfection) and incubate for 24–96 h at 37°C/5% CO_2 2. Lyse cells for SDS gel electrophoresis 3. Analyze levels of protein of interest, relative to β-actin by Western blot
Inhibition of autophagy— conversion of LC3 I to LC3 II	Western blot	1. Following transfection, seed cells in 12-well plates (1 mL per well, 4 wells per transfection) and incubate for 24–96 h at 37°C/5% CO_2 2. Treat cells ± an autophagy inducer (e.g., rapamycin) ± an autophagic flux inhibitor (e.g., bafilomycin) 3. Lyse cells for SDS gel electrophoresis 4. Analyze levels of LC3 I and LC3 II, relative to β-actin by Western blot [5] (see **Note 14**)
Inhibition of autophagy— formation of autophagosomes	Immunofluorescence/ confocal microscopy	1. Following transfection, seed cells in 12-well plates (1 mL per well, 1 well per transfection) and incubate for 24–96 h at 37°C/5% CO_2 2. Fix cells and stain with antibody against LC3, followed by fluorescently tagged secondary antibody (see Chapter 7) 3. Analyze autophagosome formation

3.2.3 LPS-Induced Autophagy

1. If required, add inhibitors to cells 15–30 min prior to addition of LPS (see Subheading 3.1).

2. Add LPS at a final concentration of 1–100 nM.

3. Incubate for 1–24 h at 37 °C/5% CO_2 (see **Note 11**).

3.3 Inhibiting Autophagy with PI3K Inhibitors

Autophagy can be inhibited by treating cells with the PI3K inhibitor 3-MA, which preferentially inhibits type III PI3K (vps34) (see Table 1). Alternatively, other pan-PI3K inhibitors, such as wortmannin and LY294002, can be used, although these may have confounding off-target effects, depending on the experimental conditions (see **Note 12**).

1. Add 3-MA (1–10 mM), wortmannin (0.1–10 µM), or LY294002 (1–100 µM) to prepared cells (see **Note 13** and Table 1).

2. Incubate for 30 min at 37 °C/5% CO_2.

3. Add further treatment, as required (e.g., recombinant MIF, MIF inhibitors, autophagy inducers, autophagic flux inhibitor).

3.4 Inhibiting Autophagy with siRNA

Small interfering RNAs (siRNA) are useful tools for manipulating autophagy genetically by targeting specific gene products on which autophagy relies. We have previously used siRNA against beclin 1 (Atg6), Atg5 and Atg7 [7, 16, 22, 23]. Here we describe two protocols, one using a transfection reagent and the other nucleofection.

3.4.1 Transfection of siRNA Using a Transfection Reagent

1. Seed cells in appropriate culture plate for desired readout, e.g., 96-well plates (flat bottomed, 100–200 μL per well) for ELISA or flow cytometry (*see also* Chapters 4–6), 12-well plates with coverslips for immunofluorescence/confocal (0.5–1 mL per well, *see also* Chapters 7 and 8), and 12- or 6-well plates (1 mL and 3 mL per well, respectively) for Western blot/co-IP. Seed at $1–3 \times 10^5$ cells/mL to achieve a final confluence of 30–50% at time of transfection.

2. Incubate for 4–24 h at 37 °C/5% CO_2.

3. Prepare siRNA in Opti-MEM. The amount required varies with the surface area of the wells: For each well of a 96-well plate, dilute 5 pmol siRNA in 25 μL Opti-MEM, for 12-well plates dilute 40 pmol siRNA in 100 μL Opti-MEM, and for 6-well plates dilute 100 pmol siRNA in 250 μL Opti-MEM. Mix gently.

4. Dilute Lipofectamine in Opti-MEM; for each well of a 96-well plate dilute 0.25 μL Lipofectamine in 25 μL Opti-MEM, for 12-well plates dilute 2 μL Lipofectamine in 100 μL Opti-MEM, for 12-well plates dilute 5 μL Lipofectamine in 250 μL Opti-MEM. Mix gently and incubate for 5 min at room temperature.

5. Combine the diluted siRNA with the diluted Lipofectamine, mix gently, and incubate for 20 min at room temperature.

6. Add the siRNA-Lipofectamine solution to the cells (total volume of 50 μL for wells of a 96-well plate, 200 μL for 12-well plates, and 500 μL for 6-well plates).

7. Incubate for 4 h at 37 °C/5% CO_2, then replace the medium with original culture medium (e.g., RPMI-1640, DMEM).

8. Incubate for 24–96 h at 37 °C/5% CO_2.

3.4.2 Nucleofection of siRNA

The following protocol is based on the Amaxa nucleofection system (Lonza), but is general enough to be easily modified for other similar systems. The Amaxa system includes specific nucleofection buffers for different cell types, each with its own specific protocol. We have found, with macrophage cells lines, that other proprietary electroporation/nucleofection buffers can also be used, but the user should optimize if using a non-Amaxa reagent.

1. Grow cells to confluence in tissue culture flasks.

2. Remove cells by scraping or with chemical disruption (e.g., trypsin). Centrifuge the cells at $300 \times g$ for 5 min.

3. Resuspend cells in 10 mL and count.

4. Centrifuge at $300 \times g$ for 5 min.

5. Resuspend cells in electroporation buffer; 2×10^6 cells in 100 μL per transfection. Transfer to 0.5 mL tubes.

6. For each transfection, add 30–400 nM siRNA (*see* **Note** 7). For each experiment, include non-targeting ("scrambled") siRNA as a control for nonspecific effects.

7. Transfer the cells/siRNA to a nucleofection cuvette. Ensure the cell suspension covers the base of the cuvette and that there are no air bubbles, as these can interfere with the electroporation.

8. Insert the cuvette into the nucleofection device and apply the relevant program, depending on cell type being transfected.

9. Add 0.5 mL of complete culture medium to the cells in the cuvette and mix gently with a sterile transfer pipette (usually supplied with the cuvettes). Transfer to final volume of medium (based on a final concentration of $2–5 \times 10^5$ cells/ mL), either in a 15-mL tube (to be transferred to culture plates later) or directly into tissue culture plates.

10. Incubate for 24–96 h at 37 °C/5% CO_2.

3.5 Verifying siRNA-Targeted Knockdown of Genes and Protein

For every experiment with siRNA, it is important to include samples for verification that the knockdown has worked. This should always be done at the protein level and can also be confirmed by polymerase chain reaction (PCR) to show knockdown of the gene. In addition, when using siRNA to target autophagy genes, it is advisable to confirm functional effects (i.e., inhibition of autophagy). For each siRNA experiment, set up one or more of the control experiments outlined in Table 2.

4   Notes

1. A number of the compounds discussed here are initially dissolved in DMSO. It is important to ensure that vehicle controls (DMSO only, at the same final concentrations) are run for all experiments using these solutions.

2. E64d is a cell-permeable ethyl ester of E-64-c. The latter is not cell-permeable, so should not be used here.

3. Both EBSS and HBSS are available from a number of companies as premixed solutions, including versions with phenol red, which is useful for assessing pH changes in the culture.

4. Different LPS serotypes are available, including O111:B4 and O55:B5 from *E. coli* and O127:B8 from *Salmonella typhimurium*. All induce autophagy well in our hands. We recommend using ultra-pure LPS, as it is less likely to contain other autophagy-regulating contaminants.

5. The 3-MA will quickly come out of solution on cooling, so use quickly and return to heat if it does.

6. Alternatively, 3-MA can be stably dissolved at 100 mM in DMSO.

7. A number of different companies offer siRNA, including many pre-designed siRNA for a wide range of autophagy genes. In addition, siRNAs are often offered as "pools" of 3 or more siRNAs against different sites on the same gene of interest, offering more potent knockdown with reduced off-target effects. We have found these to be particularly effective. However, efficiency of siRNA is variable and may differ between cell types, so the user should optimize concentrations for each siRNA and cell type.

8. A number of different transfection reagents are available from a number of different companies. Refer to manufacturer's protocols for other reagents.

9. The Amaxa system includes specific nucleofection buffers for different cell types, each with its own specific protocol. We have found, with macrophage cells lines, that other proprietary electroporation/nucleofection buffers can also be used, but the user should optimize if using a non-Amaxa reagent.

10. Cuvettes are available from a number of different companies, but it is essential to verify they work consistently in your system of choice.

11. In our hands, LPS induces autophagy in a biphasic manner; early autophagosome formation is seen in the first 2 h, which often reduces over the next few hours. A second peak, much higher than the early phase, is seen after overnight (16–24 h) after treatment (Fig. 1). This second wave of autophagy is likely driven by multiple factors released by the cells (pro-inflammatory cytokines in particular), coupled with large-scale metabolic changes. One study has shown that LPS can inhibit the maturation of phagosomes in dendritic cells [24]; it is yet to be determined whether the same is true for autophagosomes.

12. Wortmannin and LY294002 can have off-target, sometimes conflicting, effects on autophagy. They tend to work better at

shorter time point s, but it is important to confirm effects using other more specific methods, such as siRNA knockdown of autophagy genes.

13. Concentrations and timing for each inhibitor/inducer may vary with cell type, so we recommend optimizing.

14. Levels of p62/SQSTM1, which is degraded in autophagosomes, can also be measured by Western blot as a proxy for autophagy upregulation and autophagosome maturation [5].

Acknowledgements

We would like to thank Jurie Tashkandi for assistance with the experiment shown in Fig. 1.

References

1. Yin Z, Pascual C, Klionsky DJ (2016) Autophagy: machinery and regulation. Microb Cell 3 (12):588–596

2. Kaushik S, Cuervo AM (2012) Chaperone-mediated autophagy: a unique way to enter the lysosome world. Trends Cell Biol 22 (8):407–417

3. Li WW, Li J, Bao JK (2012) Microautophagy: lesser-known self-eating. Cell Mol Life Sci 69 (7):1125–1136

4. Deretic V, Levine B (2018) Autophagy balances inflammation in innate immunity. Autophagy 14(2):243–251

5. Klionsky DJ et al (2016) Guidelines for the use and interpretation of assays for monitoring autophagy (3rd edition). Autophagy 12 (1):1–222

6. Harris J (2011) Autophagy and cytokines. Cytokine 56(2):140–144

7. Peral de Castro C et al (2012) Autophagy regulates IL-23 secretion and innate T cell responses through effects on IL-1 secretion. J Immunol 189(8):4144–4153

8. Chao CH et al (2018) Macrophage migration inhibitory factor-induced autophagy contributes to thrombin-triggered endothelial hyperpermeability in sepsis. Shock 50(1):103–111

9. Chuang YC et al (2012) Macrophage migration inhibitory factor induces autophagy via reactive oxygen species generation. PLoS One 7(5):e37613

10. Xia W, Hou M (2016) Macrophage migration inhibitory factor induces autophagy to resist hypoxia/serum deprivation-induced apoptosis via the AMP-activated protein kinase/ mammalian target of rapamycin signaling pathway. Mol Med Rep 13(3):2619–2626

11. Xu S et al (2016) Macrophage migration inhibitory factor enhances autophagy by regulating ROCK1 activity and contributes to the escape of dendritic cell surveillance in glioblastoma. Int J Oncol 49(5):2105–2115

12. Xu X, Bucala R, Ren J (2013) Macrophage migration inhibitory factor deficiency augments doxorubicin-induced cardiomyopathy. J Am Heart Assoc 2(6):e000439

13. Wu MY et al (2012) Steroid receptor coactivator 3 regulates autophagy in breast cancer cells through macrophage migration inhibitory factor. Cell Res 22(6):1003–1021

14. Liu Y et al (2014) A novel androstenedione derivative induces ROS-mediated autophagy and attenuates drug resistance in osteosarcoma by inhibiting macrophage migration inhibitory factor (MIF). Cell Death Dis 5:e1361

15. Xu X, Ren J (2015) Macrophage migration inhibitory factor (MIF) knockout preserves cardiac homeostasis through alleviating Akt-mediated myocardial autophagy suppression in high-fat diet-induced obesity. Int J Obes 39(3):387–396

16. Lee JP et al (2016) Loss of autophagy enhances MIF/macrophage migration inhibitory factor release by macrophages. Autophagy 12 (6):907–916

17. Ichimura Y et al (2000) A ubiquitin-like system mediates protein lipidation. Nature 408 (6811):488–492

18. Harris J, Hanrahan O, De Haro SA (2009) Measuring autophagy in macrophages. Curr Protoc Immunol Chapter 14: p. Unit 14 14

19. Mizushima N, Yoshimori T (2007) How to interpret LC3 immunoblotting. Autophagy 3 (6):542–545

20. Shi CS, Kehrl JH (2010) TRAF6 and A20 regulate lysine 63-linked ubiquitination of Beclin-1 to control TLR4-induced autophagy. Sci Signal 3(123):ra42

21. Wang J et al (2013) Lipopolysaccharide (LPS)-induced autophagy is involved in the restriction of Escherichia coli in peritoneal mesothelial cells. BMC Microbiol 13:255

22. Harris J et al (2007) T helper 2 cytokines inhibit autophagic control of intracellular Mycobacterium tuberculosis. Immunity 27 (3):505–517

23. Harris J et al (2011) Autophagy controls IL-1beta secretion by targeting pro-IL-1beta for degradation. J Biol Chem 286 (11):9587–9597

24. Alloatti, A et al. (2015). Toll-like Receptor 4 engagement on dendritic cells restrains phago-Lysosome fusion and promotes cross-Presentation of antigens. Immunity 43(6):1087–1100

Chapter 14

Assays for Measuring the Role of MIF in NLRP3 Inflammasome Activation

Anita A. Pinar and James Harris

Abstract

Hallmarks of NLRP3 inflammasome activation include the cleavage and secretion of the mature forms of caspase-1, IL-1β, and IL-18 and aggregation of ASC into "specks." We have previously shown that macrophage migratory inhibitory factor (MIF) directly regulates activation of the NLRP3 inflammasome, inhibiting the release of interleukin (IL)-1α, IL-1β, and IL-18. Here we present protocols for studying activation of the NLRP3 inflammasome in human and mouse macrophages and peripheral blood mononuclear cells (PBMCs). These protocols can also be applied to different cell types, such as fibroblasts, neutrophils, endothelial cells, and epithelial cells, although further optimization may be required for each. We also cover the stimulation of macrophages with established NLRP3 inflammasome activators.

Key words MIF, NLRP3, ELISA, Immunoblotting, ASC oligomerization, Caspase-1 activity, Interleukin-1α activation, Interleukin-1β activation, Interleukin-18 activation, NLRP3 reporter assays

1 Introduction

We and others have recently demonstrated that MIF facilitates the release of IL-1α, IL-1β, and IL-18 via activation of the NLRP3 inflammasome [1, 2]. In macrophages, we have found this to be a reliable and consistent biological role for MIF and, as a result, assays measuring NLRP3 activation and its consequences represent potentially useful indicators of MIF activity [1]. Activation of the NLRP3 inflammasome results in the recruitment of the adaptor protein ASC (apoptosis-associated speck-like protein containing a CARD), which is required for the cleavage of pro-caspase-1 into the active enzyme, caspase-1. In the absence of inflammasome activators, ASC has a diffuse distribution in the cytoplasm and nucleus of cells (Fig. 1). In response to stimulation with NLRP3 inflammasome activators, such as nigericin [3], ATP [3], silica crystals [4], MSU crystals [5], cholesterol crystals [6], influenza A PB1-F2 peptide [7–9], IAPP (amylin, or islet amyloid peptide)

James Harris and Eric F. Morand (eds.), *Macrophage Migration Inhibitory Factor: Methods and Protocols*,
Methods in Molecular Biology, vol. 2080, https://doi.org/10.1007/978-1-4939-9936-1_14,
© Springer Science+Business Media, LLC, part of Springer Nature 2020

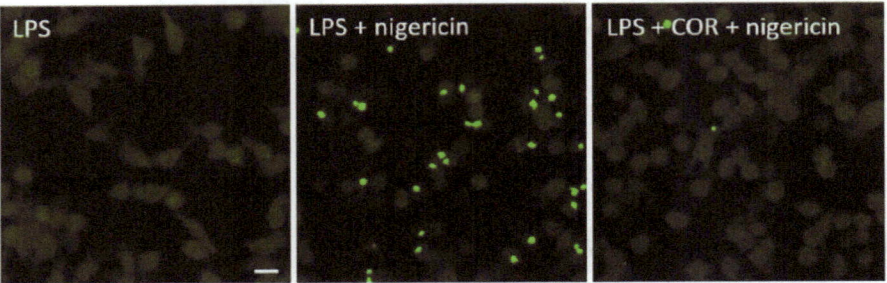

Fig. 1 ASC "speck" assay. In the absence of inflammasome activators such as nigericin, ASC adaptor proteins in LPS-treated ASC-cerulean iBMMs (*left panel*) have a dull and diffuse distribution, localized to the cytoplasm and nucleus with an intact and rounded cell morphology. Treatment of ASC-cerulean iBMMs with the NLRP3 inflammasome-specific activator nigericin rapidly promotes oligomerization of ASC proteins into a large perinuclear "speck" (*middle panel*). Many of these cells also have irregular cell morphology, as they are undergoing pyroptosis. Treatment with the MIF inhibitor, COR123625 inhibits this process (*right panel*)

[10], imiquimod [11], and high salt [12], ASC proteins oligomerize into large perinuclear "specks" (Fig. 1). Typically, cells contain a single ASC speck following inflammasome activation, which is large enough to be visualized using fluorescence microscopy and is a useful readout for inflammasome activation upstream of caspase-1 activation [1]. Additional readouts include the detection of mature caspase-1, IL-1β, and IL-18 via immunoblotting (Fig. 2), secretion of IL-1α, IL-1β, and IL-18 by ELISA, and caspase-1-dependent cell death pyroptosis by LDH detection. It is important to note that chemical activators such as nigericin activate the NLRP3 inflammasome via a different mechanism compared to a particulate activator such as silica, which takes more time to induce activation of the inflammasome. Consequently, the timing and dosages of these and other activators differ and need to be noted for capturing the inflammasome activation events, as indicated throughout the protocols in this chapter.

Here, we detail assays for stimulating NLRP3 activation in macrophages and measuring a number of different outputs. ASC speck formation is measured by immunofluorescence/confocal microscopy and flow cytometry, the release of IL-1 family protein is measured using ELISA and cell reporter (HEK-Blue) assays, levels and processing of inflammasome proteins is analyzed by Western blot, caspase-1 activity by FLICA assay, and cell death (pyroptosis) by LDH assay.

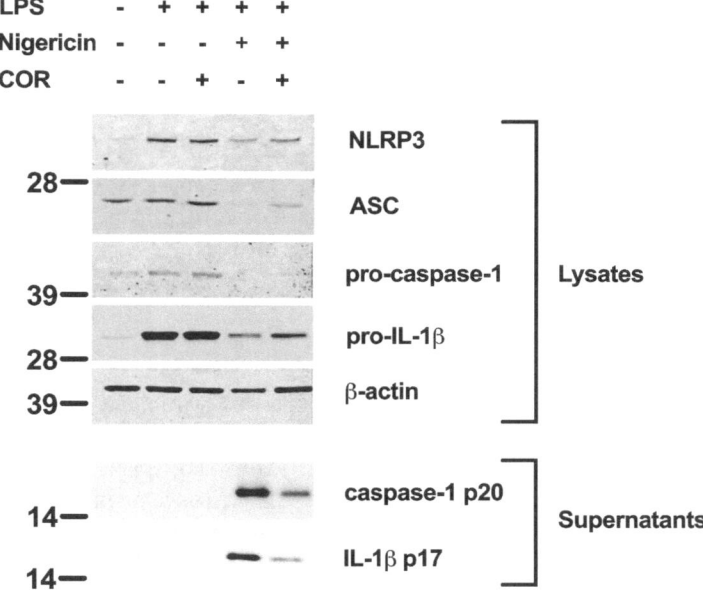

Fig. 2 Immunoblotting cell lysates and culture supernatants for NLRP3 inflammasome activation. Wild-type bone marrow-derived macrophages were left untreated, primed with LPS alone (100 ng/mL) for 5 h, pretreated with MIF inhibitor COR123625 (50 μM) for 1 h prior to the addition of LPS, primed with LPS and treated with nigericin (5 μM) or treated with COR123625, primed with LPS and treated with nigericin. Western blot analysis of cellular lysates and supernatants to assess levels of NLRP3, ASC, pro-IL-1β, pro-caspase-1, IL-1β p17, caspase-1 p20, and β-actin. Taken from [1]. Reproduced under a Creative Commons Attribution 4.0 International licence (http:/creativecommons.org/licenses/by/4.0/)

2   Materials

2.1   Culturing Macrophages

1. Macrophages expressing fluorescently tagged ASC, e.g., ASC-cerulean immortalized bone marrow-derived macrophages (iBMM).

2. Differentiated bone marrow-derived macrophages (BMM) isolated from WT and $Mif^{-/-}$ mice.

3. Complete RPMI medium: RPMI-1640 medium supplemented with 10% FCS, 2 mM L-glutamine, 50 U/mL penicillin, and 50 μg/mL streptomycin.

4. Serum-free RPMI medium: RPMI-1640 medium supplemented with 2 mM L-glutamine and 1% penicillin/streptomycin.

5. 6-Well, 12-well, 96-well tissue culture plates.

2.2 Stimulation of Macrophages

1. Complete RPMI medium: RPMI 1640 supplemented with 10% fetal calf serum (FCS), 2 mM L-glutamine, 50 U/mL penicillin, and 50 µg/mL streptomycin.

2. Incubator maintained at 37 °C in a humidified atmosphere of 5% CO_2.

3. Ultra-pure LPS (stock solution of 1 mg/mL in ultra-pure water, final concentrations of 0.05–100 ng/mL in serum-free RPMI) (*see* **Note 1**).

4. MIF inhibitor (appropriate stock solution in serum-free RPMI) (*see* **Note 2**).

5. Transfection reagent (*see* **Note 3**).

6. NLRP3 inflammasome activators (all in serum-free RPMI) (*see* **Note 4**). Examples include: nigericin (5 mM for a final concentration of 5–10 µM); ATP (5–10 mM); silica crystals (150 µg/mL); monosodium urate crystals (MSU; 150 µg/mL); cholesterol crystals (250 µg/mL); PB1-F2 peptide (100 µg/mL); islet amyloid particulate protein (IAPP; 10 µM); R837 (Imiquimod; 20 µg/mL).

7. Other inflammasome activators as negative controls (in serum-free RPMI). Examples include: Poly dA:dT (1 µg/mL; AIM2 inflammasome activator) and flagellin (250 ng/mL; NLRC4 activator).

2.3 ELISA for Detection of Secreted IL-1α, IL-1β, and IL-18

1. Human/murine IL-1α ELISA kits.

2. Human/murine IL-1β ELISA kits.

3. Human/murine IL-18 ELISA kits.

4. TMB Substrate Solution (if not provided in the ELISA kit).

5. Stop Solution: 0.5 M H_2SO_4 (if not provided in the ELISA kit).

6. Microplate reader capable of measuring absorbance at 450 nm (and, if possible, 540 or 570 nm for reference).

7. Pipettes and pipette tips.

8. Distilled water.

2.4 Isolation of Protein Fraction from Supernatants

1. Supernatant samples from stimulated macrophages.

2. 1× Sample buffer: 25% Glycerol, 2% SDS, 50 mM Tris–HCl (pH 6.8), 0.1% bromophenol blue, 100 mM DTT (1,4-dithiothreitol, reducing agent, added fresh) in milliQ water.

3. Methanol.

4. Chloroform.

5. Heat block.

2.5 Cell Lysis for Immunoblotting

1. RIPA lysis buffer: 2 mM Tris–HCl (pH 7.5), 150 mM NaCl, 1 mM Na$_2$EDTA, 1 mM EGTA, 1% NP-40, 1% sodium deoxycholate, 2.5 mM sodium pyrophosphate, 1 mM β-glycerophosphate, 1 mM Na$_3$VO$_4$, 1 μg/mL leupeptin (*see* **Note 5**).

2. Ice.

3. Cell scrapers.

4. Microcentrifuge.

5. 2× Sample buffer: 50% Glycerol, 4% SDS, 100 mM Tris-Cl (pH 6.8), 0.2% bromophenol blue, 200 mM DTT (1,4-dithiothreitol, reducing agent, added fresh) in milliQ water.

2.6 Immunoblotting

1. SDS PAGE Gel Running System

2. Protein transfer system.

3. PVDF or nitrocellulose protein transfer membrane.

4. Ponceau S red stain: 0.1% (w/v) Ponceau S in 5% (v/v) acetic acid.

5. Tris-buffered saline (TBS): 50 mM Tris-Cl, 150 mM NaCl. Adjust pH to 7.6 with 1 M HCl.

6. TBS with Tween-20 (TBS-T): TBS with 0.05% Tween-20 (polyoxyethylene sorbitane monolaureate).

7. Immunoblotting blocking solution: 5% w/v skim milk in TBS-T (*see* **Note 6**).

8. Anti-NLRP3 antibody.

9. Anti-ASC antibody.

10. Anti-caspase-1 antibody (*see* **Note 7**).

11. Anti-IL-1β antibody.

12. Anti-MIF antibody.

13. HRP-conjugated secondary antibodies.

14. Western blot ECL substrate.

15. X-ray film or chemiluminescence detection system.

2.7 ASC Speck Assay in iBMM Expressing Fluorescently Tagged ASC

1. iBMM expressing fluorescently tagged ASC (e.g., ASC-cerulean or ASC-mCherry).

2. Complete RPMI medium.

3. 12-Well tissue culture plates.

4. Round glass coverslips (#1.5; 17 mm diameter).

5. Microscope slides.

6. Ultra-pure LPS stock solution (1 mg/mL).

7. MIF inhibitor (*see* **Note 2**).

8. NLRP3 inflammasome activators (*see* Subheading 2.2).

9. Ice-cold methanol or 2% PFA in phosphate-buffered saline (PBS) (*see* **Note 8**).

10. Ice-cold PBS.

11. Fluorescence mounting medium.

12. Confocal microscope equipped with a cerulean (Diode/ 440 nm) filter.

13. Imaging software.

2.8 ASC Speck Assay in Other Cells

1. Cells of choice (e.g., PBMCs, human macrophages, primary mouse macrophages).

2. 12-Well plates.

3. Glass coverslips (#1.5; 17 mm diameter).

4. Ultra-pure LPS (10–100 ng/mL).

5. MIF inhibitor (*see* **Note 2**).

6. NLRP3 inflammasome activators (*see* Subheading 2.2).

7. PBS.

8. Ice-cold methanol or 2% PFA in PBS (*see* **Note 8**).

9. Permeabilization buffer: 0.1% Triton X-100 in PBS.

10. Immunofluorescence blocking buffer: 0.5% teleostein gelatin, 1% casein, 5% serum (*see* **Note 9**) in PBS.

11. Anti-ASC antibody (1:100–1:1000) in blocking buffer.

12. Fluorophore-conjugated secondary antibodies (e.g., Alexa-Fluor®, DyLight®) (1/1000–1/2000) in blocking buffer.

13. Nuclear stain (e.g., DAPI, 1–5 mg/mL stock solution, final concentration 1–5 µg/mL).

14. Mounting medium.

15. Confocal microscope.

16. Imaging software.

2.9 Detection of IL-1β and IL-18 by HEK-Blue IL-1β IL-18 Cells

1. HEK-Blue IL-1β/HEK-Blue IL-18 cell suspensions.

2. Warmed PBS.

3. Cell scraper.

4. Microcentrifuge.

5. RPMI-1640 medium.

6. 96-Well flat-bottom plate.

7. Recombinant human IL-1β (0.25 µg/mL in PBS or complete RPMI).

8. Recombinant human TNF-α (0.25 µg/mL in PBS or complete RPMI) (*see* **Note 10**).

9. Incubator.

10. QUANTI-Blue: Pour the contents of one pouch of QUANTI-Blue™ into 100 mL of sterile water and gently swirl to mix. Warm QUANTI-Blue™ to 37 °C for 30 min (*see* **Note 11**).

11. Spectrophotometer at 620–655 nm.

2.10 Caspase-1
Activity Assay

1. Stimulated macrophages.

2. Caspase-1 activity assay kit.

3. Fluorescence microplate reader.

2.11 LDH Detection
for Cell Death
(Pyroptosis)

1. Cell and supernatant samples from stimulated macrophages.

2. LDH assay kit.

3. Fluorescence microplate reader.

3 Methods

3.1 Preparation
of Macrophages

1. Plate differentiated murine WT and *Mif*$^{-/-}$ BMMs [13] (also *see* Chapter 4), human PBMCs [14], differentiated THP1 cells, or ASC-cerulean iBMM in complete RPMI according to the assay to be performed: For ELISA: 5×10^5 cells/mL (200 μL per well) in 96-well plate format. For immunoblotting: 0.5×10^6 cells/mL (2 mL per well) in 6-well plate format. For imaging assays: 0.2×10^6 cells/mL (1 mL per well) on coverslips in 12-well plate format.

2. Allow cells to settle/adhere for 2–24 h before stimulation.

3.2 Stimulation
of Macrophages

1. Where MIF inhibition is used, treat cells with inhibitor for 1–2 h prior to treatment with other stimuli.

2. To prime cells (*see* **Note 12**), add LPS (10 ng/mL for BMMs or 50 pg/mL for PBMCs [15] for 4–24 h) (*see* **Note 13**).

3. Treat primed cells with NLRP3 inflammasome activators, e.g., nigericin (5 μM) or ATP (5–10 mM) for 30–120 min; MSU (150 μg/mL), silica (150 μg/mL), cholesterol crystals (250 μg/mL), IAPP (10 μM), or PB1-F2 peptide (100 μg/mL) for 3–6 h.

4. Centrifuge the plate for 5 min at $300 \times g$ (*see* **Note 14**).

5. Depending on the assay of choice, you may choose to: (1) collect supernatants (*see* **Note 15**) and assay for secreted cytokines (*see* Subheading 3.3), perform immunoblotting (*see* Subheadings 3.4 and 3.6) or measure LDH levels (*see* Subheading 3.11), (2) lyse cells for immunoblotting (*see* Subheadings 3.5 and 3.6), (3) prepare cells for immunofluorescence/confocal microscopy (*see* Subheadings 3.7 and 3.8), flow cytometry (*see* Subheading 3.9) or caspase-1 activity assay (*see* Subheading 3.10).

3.3 ELISA for Detection of Secreted IL-1α, IL-1β, and IL-18

1. Assay for levels of secreted IL-1α, IL-1β, and IL-18 from supernatants using ELISA kits according to manufacturer's instructions.

3.4 Isolating Protein Fraction from Culture Supernatants

1. Take supernatants from Subheading 3.2.

2. In an eppendorf tube, mix 500 μL supernatant with 500 μL methanol and 100 μL chloroform and vortex for 30 s.

3. Centrifuge for 10 min at 13,000 rpm in a microcentrifuge.

4. Discard the upper phase without disturbing the pellet at the interface.

5. Add 500 μL methanol and vortex for 30 s.

6. Centrifuge for 10 min at 13,000 rpm in a microcentrifuge to obtain a concentrated protein pellet.

7. Remove supernatant carefully and air-dry for 20 min at RT.

8. Resuspend concentrated protein in 20–40 μL of 1× sample buffer. Samples can be stored at −20 to −80 °C at this point.

9. Boil samples at 95 °C for 5–10 min. These are now ready for immunoblotting (*see* Subheading 3.6).

3.5 Preparing Cell Lysates for Immunoblotting

1. Lyse cells in 100 μL RIPA lysis buffer in ice for 30–60 min.

2. Scrape cells on ice.

3. Centrifuge lysed cells for 10 min at 300 g to remove nuclei/ large debris.

4. Collect the supernatant (lysate) to assay for ASC oligomerization and expression of NLRP3, pro-IL-1β, and pro-IL-18. This can be stored at −20 to −80 °C before use.

5. Mix lysate 1:1 with 2× sample buffer and heat at 95 °C for 5–10 min.

3.6 Immunoblotting

1. Run samples on SDS-PAGE.

2. Transfer proteins from the SDS-PAGE gel to PVDF or nitrocellulose membrane using a protein transfer system.

3. If required, incubate the membrane in Ponceau S stain to detect transferred protein bands. This staining is reversible and can be removed by repeated rinses in TBS-T.

4. Incubate the membrane in immunoblotting blocking solution on a rotor at room temperature for 1 h.

5. Incubate the membrane with the desired primary antibody diluted in immunoblotting blocking solution overnight at 4 °C.

6. Wash the membrane three times for 5 min each in TBS-T.

7. Incubate the membrane with HRP-conjugated secondary antibody diluted in immunoblotting blocking solution for 1 h at room temperature (*see* **Note 16**).

8. Visualize bands on the membrane with ECL using X-ray film or a chemiluminescence detection system (*see* **Note 16**).

9. Quantification using densitometry analysis can be performed using appropriate image analysis software (e.g., Image J/FIJI). Protein levels should be normalized to respective loading controls (e.g., β-actin, GAPDH).

3.7 Detection of ASC Speck Formation in iBMM Expressing Fluorescently Tagged ASC

1. Stimulate fluorescently tagged ASC-expressing iBMM (*see* **Note 17**), cultured on coverslips, as in Subheading 3.2.

2. Remove treatment media and fix cells in 500 μL ice-cold methanol for 5 min at −20 °C or 2% PFA for 30 min at room temperature (*see* **Note 8**).

3. Wash cells three times in ice-cold PBS.

4. Mount coverslip with adherent cells (cell side down) onto microscopy slides with fluorescence mounting medium and, if required, allow to set overnight (*see* **Note 18**).

5. Image a range of 1 μm z-stacks from multiple fields of view for subsequent z-projections using a confocal microscope equipped with a cerulean filter (*see* **Note 19**).

6. Quantify the percentage of cells with ASC specks (typically one per cell) in each field of view imaged using imaging software [16].

3.8 Immunostaining for the Detection of ASC Specks

1. Stimulate cells, grown on coverslips, as in Subheading 3.2.

2. Remove culture supernatant. Fix cells in 500 μL ice-cold methanol for 5 min at −20 °C or 2% PFA for 30 min at room temperature (*see* **Note 8**).

3. Wash cells three times using 1 mL ice-cold 1x PBS.

4. Permeabilize cells in 500 μL 0.1% Triton X-100 for 10–30 min at room temperature.

5. Wash cells three times using 1 mL ice-cold 1× PBS.

6. Incubate cells with immunofluorescence blocking buffer for 1 h at room temperature.

7. Incubate cells with primary antibody in blocking solution at room temperature for 1–2 h (*see* **Note 20**). This step can be done in the wells of the tissue culture plate, but this will require a large amount of antibody. Alternatively, 50 μL of antibody solution can be dropped on parafilm stretched over the base of a tissue culture plate and the coverslip put on this (cell side down) (*see* Chapter 8). This can be covered and incubated as above.

8. Return coverslips to the wells (cell side up). Wash cells three times using 1 mL ice-cold 1× PBS.

9. In the wells of the tissue culture plate, incubate cells with the corresponding fluorophore-labeled secondary antibodies for 1–2 h at room temperature.

10. Wash cells three times using 1 mL ice-cold 1× PBS.

11. If required, co-stain cells with nuclear stain for 5 min at RT.

12. Wash cells three times using 1 mL ice-cold 1× PBS.

13. Mount cells with fluorescence mounting medium and allow to set overnight, if necessary.

14. Image multiple 1 μm z-stacks over multiple fields of view using a confocal microscope.

15. Quantify the percentage of cells with ASC specks using imaging software.

3.9 Detection of IL-1β and IL-18 by HEK-Blue IL-1β IL-18 Cells

1. On day 1, prepare HEK-Blue IL-1β and IL-18 cell suspensions (*see* **Note 21**).

2. Wash cells twice with pre-warmed (37 °C) PBS.

3. Detach the cells in PBS using a cell scraper (*see* **Note 22**).

4. Centrifuge cells at 300 g for 10 min and resuspend cells at 3.3×10^5 cells per mL in fresh complete RPMI medium (96-well flat-bottom plate format).

5. On day 2, stimulate human macrophages/PBMC/THP-1 cells for NLRP3 activation, as in Subheading 3.2.

6. Transfer 50 μL of supernatants from stimulated cells to wells of a 96-well plate.

7. As a positive control, add 50 μL of recombinant human IL-1β (0.25 μg/mL) to one set of wells and/or 50 μL of recombinant human IL-18 (0.25 μg/mL) to another set of wells.

8. As a negative control, add 50 μL recombinant human TNF-α (0.25 μg/mL) (*see* **Note 10**) to another set of wells.

9. Add 150 μL of HEK-Blue IL-1β and IL-18 cell suspensions (50,000 total cells) per well. Incubate both overnight at 37 °C in 5% CO_2.

12. The next day, prepare QUANTI-Blue (*see* **Note 11**) and add QUANTI-Blue (150 μL/well).

13. Add 50 μL of induced HEK-Blue IL-1β and HEK-Blue IL-18 cell supernatants.

14. Incubate the IL-1β and IL-18 plates at 37 °C for 30 min to 3 h.

15. Measure the levels of SEAP (*see* **Note 23**) from both HEK-Blue IL-1β and HEK-Blue IL-18 cells using a spectrophotometer at 620–655 nm.

3.10 Caspase-1 Activity Assay

1. Stimulate cells, as in Subheading 3.2, in 96-well black plates (clear bottoms).

2. Caspase-1 activity assay kit contains a substrate YVAD which binds to active caspase-1 allowing for the assessment of levels of active caspase-1. Perform assay according to manufacturer's instructions (*see* **Note 24**).

3.11 LDH Detection for Extent of Pyroptosis

1. Stimulate cells, as in Subheading 3.2, in 96-well flat-bottom plates (clear).

2. Perform assay with supernatant samples from stimulated macrophages according to manufacturer's instructions and then calculate relative cytotoxicity (*see* **Note 25**).

4 Notes

1. Different LPS serotypes are available, including O111:B4 and O55:B5 from *E. coli* and O127:B8 from *Salmonella typhimurium*. All are capable of priming inflammasome activation. However, O111:B4 is also capable of inducing noncanonical inflammasome activation in concert with other stimuli [17], so care should be taken when interpreting results using this serotype.

2. A number of different MIF inhibitors are commercially available, including ISO-1, 4-IPP, p425, and RDR 03785. These inhibitors may work in different ways. We have shown inhibition of NLRP3 activation with ISO-1 and 4-IPP, but have not tested the others. We recommend using more than one inhibitor and optimizing conditions for each. Note also that MIF-neutralizing antibodies could also be tested here.

3. Cationic-lipid transfection reagents are available from a number of different manufacturers. We have no reason to believe any of these are not suitable, but recommend optimization for each.

4. The exact mechanisms for how different NLRP3 inflammasome activators induce activation are still under investigation. However, what is observed is that the number of ASC speck-positive cells increases over time with increased cell death; therefore it is important to optimize the timing of stimulation for each NLRP3 inflammasome activator used. Chemical activators, such as nigericin, typically take a shorter time (up to 1 h) compared to particulate activators, such as silica, which can take 4–6 h to induce activation.

5. Other lysis buffers may be preferable in some instances; we recommend optimisation for each assay.

6. Alternative blocking solutions can also be used including 1–5% BSA and fish gelatin.

7. Different caspase-1 antibodies to detect the p10 vs. p20 can be used [18].

8. It is advised to try different fixation protocols for different situations/conditions. We have found that methanol is not appropriate for some applications/fluorescent tags, while some antibodies may work better with one particular fixation technique.

9. The serum used here should be the same species as the secondary antibody used, e.g., use goat serum when using a goat anti-mouse/rabbit conjugated secondary antibody.

10. HEK-Blue IL-1β cells should not respond to human TNF-α.

11. QUANTI-Blue is a colorimetric enzyme assay for quantifying alkaline phosphatase activity. Reconstituted QUANTI-Blue solution is an effective means of detecting and quantifying SEAP and AP activity. Reconstituted QUANTI-Blue can be aliquotted and stored at 4 °C.

12. BMMs require a priming step in order to induce expression of pro-IL-1β, pro-IL-18, and inflammasome components, such as NLRP3, ASC, and pro-caspase-1. LPS is commonly used, but other TLR agonists can also prime macrophages.

13. For some assays, it may be necessary or helpful to replace the supernatant at this stage with serum-free RPMI (supplemented with L-glutamine and penicillin/streptomycin). This can help reduce unwanted protein from serum in immunoblot assays, for example. However, we advise performing experiments with and without this media replacement, as culturing cells in serum-free media may alter their responses.

14. This is to ensure that treatment particulates and cell debris do not interfere with the IL-1α, IL-1β, and IL-18 ELISA assays. This is especially important if samples are to be frozen before analysis.

15. The supernatant samples contain the processed and active forms of IL-1α, IL-1β, and IL-18, whereas the cell lysates contain the unprocessed and pro-forms of these inflammasome-induced cytokines.

16. Immunoblots can also be imaged using fluorophore-conjugated secondary antibodies (e.g., Alexa-Fluor®, DyLight®) and a fluorescent imaging system.

17. There are a variety of fluorescently tagged ASC constructs available, including ASC-cerulean-/citrine-/mCherry tagged iBMM.

18. Some mounting media form a solid polymer, while others remain in fluid phase. For the latter, the coverslip needs to be sealed around the edges to hold it in place. This can be done with nail varnish.

19. To ensure that the confocal recording and detection of ASC specks are maximized.

20. Alternatively, cells can be incubated with primary antibody overnight at 4 °C.

21. HEK-Blue™ IL-1β and HEK-Blue™ IL-18 cells allow for the detection of bioactive IL-1β and IL-18, respectively, by monitoring activation of the NF-κB and AP-1 pathways. They express a NF-κB/AP-1 inducible SEAP reporter gene. Binding of IL-1β or IL-18 to their respective receptors, IL-1R or IL-18R, on the surface of HEK-Blue™ IL-1β or HEK-Blue™ IL-18 cells triggers a signaling cascade leading to the activation of NF-κB/AP-1 and the subsequent production of SEAP.

22. Do not use trypsin to detach HEK-Blue-IL-1β cells as trypsin can alter their response.

23. Secreted embryonic alkaline phosphatase (SEAP) is used to study promoter activity or gene expression. As SEAP is secreted into cell culture supernatant, it allows for ease of determining NF-κB and AP-1 reporter activity.

24. The assay is based on detection of the cleaved substrate YVAD-AFC. Comparison of the fluorescence of AFC from a treated sample in comparison to an untreated control allows for the determination of the fold increase in caspase-1 activity.

25. To calculate relative cytotoxicity, five control values are required: (1) effector cell spontaneous LDH release control value—corrects for spontaneous release of LDH from effector cells; (2) target cell spontaneous LDH release control value—corrects for spontaneous release from target cells; (3) target cell maximum LDH release control value; (4) volume correction control value—corrects the volume increase caused by the increase in volume by the 10× lysis buffer, and (5) the culture medium background control—corrects for any LDH activity that may be present in serum containing culture medium. Compute values (1)–(5) into the below formula to calculate relative cytotoxicity according to the formula below:

$$\text{Relative Cytotoxicity} = \frac{\text{Experimental value} - \text{Effector cells spontaneous control} - \text{Target cells spontaneous control}}{\text{Target cell maximum control} - \text{Target cells spontaenous control}} \times 100$$

Acknowledgements

We would like to thank Tali Lang for reviewing this chapter and providing invaluable input.

References

1. Lang T, Lee JPW, Elgass K et al (2018) Macrophage migration inhibitory factor is required for NLRP3 inflammasome activation. Nat Commun 9(1):2223. https://doi.org/10.1038/s41467-018-04581-2

2. Shin MS, Kang Y, Wahl ER et al (2019) Macrophage migration inhibitory factor regulates U1 small nuclear RNP immune complex-mediated activation of the NLRP3 inflammasome. Arthritis Rheumatol (Hoboken, NJ) 71 (1):109–120. https://doi.org/10.1002/art.40672

3. Muñoz-Planillo R, Kuffa P, Martínez-Colón G et al (2013) K(+) efflux is the common trigger of NLRP3 inflammasome Activation by bacterial toxins and particulate matter. Immunity 38 (6):1142–1153. https://doi.org/10.1016/j.immuni.2013.05.016

4. Dostert C, Petrilli V, Van Bruggen R et al (2008) Innate immune activation through Nalp3 inflammasome sensing of asbestos and silica. Science (New York, NY) 320 (5876):674–677. https://doi.org/10.1126/science.1156995

5. Martinon F, Petrilli V, Mayor A et al (2006) Gout-associated uric acid crystals activate the NALP3 inflammasome. Nature 440 (7081):237–241. https://doi.org/10.1038/nature04516

6. Duewell P, Kono H, Rayner KJ et al (2010) NLRP3 inflammasomes are required for atherogenesis and activated by cholesterol crystals. Nature 464(7293):1357–1361. https://doi.org/10.1038/nature08938

7. McAuley JL, Tate MD, MacKenzie-Kludas CJ et al (2013) Activation of the NLRP3 inflammasome by IAV virulence protein PB1-F2 contributes to severe pathophysiology and disease. PLoS Pathog 9(5):e1003392. https://doi.org/10.1371/journal.ppat.1003392

8. Pinar A, Dowling JK, Bitto NJ et al (2017) PB1-F2 Peptide Derived from avian influenza A virus H7N9 Induces inflammation via activation of the NLRP3 inflammasome. J Biol Chem 292(3):826–836. https://doi.org/10.1074/jbc.M116.756379

9. Alvarado R, To J, Lund ME et al (2017) The immune modulatory peptide FhHDM-1 secreted by the helminth Fasciola hepatica prevents NLRP3 inflammasome activation by inhibiting endolysosomal acidification in macrophages. FASEB J 31(1):85–95. https://doi.org/10.1096/fj.201500093R

10. Masters SL, Dunne A, Subramanian SL et al (2010) Activation of the NLRP3 inflammasome by islet amyloid polypeptide provides a mechanism for enhanced IL-1beta in type 2 diabetes. Nat Immunol 11(10):897–904. https://doi.org/10.1038/ni.1935

11. Gross CJ, Mishra R, Schneider KS et al (2016) K(+) efflux-independent NLRP3 inflammasome activation by small molecules targeting mitochondria. Immunity 45(4):761–773. https://doi.org/10.1016/j.immuni.2016.08.010

12. Krishnan SM, Dowling JK, Ling YH et al (2016) Inflammasome activity is essential for one kidney/deoxycorticosterone acetate/salt-induced hypertension in mice. Br J Pharmacol 173(4):752–765. https://doi.org/10.1111/bph.13230

13. Weischenfeldt J, Porse B (2008) Bone marrow-derived macrophages (BMM): isolation and applications. CSH Protoc 2008:pdb prot5080. https://doi.org/10.1101/pdb.prot5080

14. Dagur PK, McCoy JP (2015) Collection, storage, and preparation of human blood cells. Curr Protoc Cytom 73:5.1.1–5.1.16. https://doi.org/10.1002/0471142956.cy0501s73

15. Mizote Y, Wakamatsu K, Ito S et al (2014) TLR4 and NLRP3 inflammasome activation in monocytes by N-propionyl cysteaminylphenol-maleimide-dextran (NPCMD). J Dermatol Sci 73(3):209–215. https://doi.org/10.1016/j.jdermsci.2013.11.006

16. Fernandes-Alnemri T, Wu J, Yu JW et al (2007) The pyroptosome: a supramolecular assembly of ASC dimers mediating inflammatory cell death via caspase-1 activation. Cell Death Differ 14(9):1590–1604. https://doi.org/10.1038/sj.cdd.4402194

17. Kayagaki N, Wong MT, Stowe IB et al (2013) Noncanonical inflammasome activation by intracellular LPS independent of TLR4. Science (New York, NY) 341(6151):1246–1249. https://doi.org/10.1126/science.1240248

18. Boucher D, Monteleone M, Coll RC et al (2018) Caspase-1 self-cleavage is an intrinsic mechanism to terminate inflammasome activity. J Exp Med 215(3):827–840. https://doi.org/10.1084/jem.20172222

Chapter 15

Assays for Inducing and Measuring Cell Death to Detect Macrophage Migration Inhibitory Factor (MIF) Release

Shahrzad Zamani, Eric F. Morand, and Jacqueline K. Flynn

Abstract

Cell death is a vital process for maintaining tissue homeostasis and removing potentially harmful cells. Cell death can be both programmed and non-programmed and is commonly divided into two main forms, termed apoptotic and necrotic death modes. In this chapter cell death is classified into apoptosis, primary necrosis, pyroptosis, and necroptosis. This chapter outlines the measurement of these different types of cell death and the relationship of measuring MIF release in these assays.

Key words MIF, Apoptosis, Necrosis, Pyroptosis, Necroptosis, Flow cytometry, Colorimetric assay

1 Introduction

Macrophage migration inhibitory factor (MIF) is an immunoregulatory cytokine and pleiotropic inflammatory mediator that is constitutively produced by a variety of cells, including monocytes and macrophages [1–3]. In contrast to many cytokines, MIF is constitutively expressed and stored in intracellular cytoplasmic pools, and therefore does not require de novo protein synthesis before secretion. As a result, MIF can be rapidly released in response to stimuli, such as microbial products, pro-inflammatory mediators, proliferative signals, hypoxia, and stress [3, 4]. Importantly, MIF is implicated in the pathogenesis of sepsis [5] and inflammatory and autoimmune diseases such as rheumatoid arthritis [6, 7] and systemic lupus erythematosus [8, 9], thus suggesting that MIF-directed therapies might offer new treatment opportunities for human diseases in the future [10]. Multiple clinical studies have also pointed to the potential of MIF as a biomarker in the context of inflammatory diseases, for example, systemic infections and sepsis, autoimmune diseases, cancer, and metabolic disorders [11].

The mechanism by which MIF is secreted/released has not been fully established, although in a recent study MIF was found

James Harris and Eric F. Morand (eds.), *Macrophage Migration Inhibitory Factor: Methods and Protocols*,
Methods in Molecular Biology, vol. 2080, https://doi.org/10.1007/978-1-4939-9936-1_15,
© Springer Science+Business Media, LLC, part of Springer Nature 2020

Table 1
Methods for the induction of cell death

Cell death induced	Inducer/s	Intermediate signalling	Inhibitor/s	Detection methods	References
Apoptosis	Staurosporine	Caspase dependent	Pan caspase inhibitor (Z-VAD-FMK)	Flow cytometry assay: quantification of Annexin V and PI staining Early apoptotic cells are Annexin V positive and PI negative	[18–20]
Necrosis	Ethanol	None	Pan caspase inhibitor (Z-VAD-FMK)	Flow cytometry assay: quantification of Annexin V and PI staining Late apoptotic/necrotic cells are Annexin V positive and PI positive	[21]
Pyroptosis	LPS + Nigericin	Caspase-1 dependent, NLRP3 inflammasome activation	Pan caspase inhibitor (Z-VAD-FMK) Caspase-1 inhibitor (Z-YVAD-FMK)	Florescence-based detection of LDH activity MIF and IL-1β ELISA	[29]
Necroptosis	TNF-α + Z-VAD-FMK + BV-6	Caspase independent, RIP1 kinase	Necrostatin-1	Florescence-based detection of LDH activity MIF ELISA	[31]

LPS lipopolysaccharide, *TNF* tumor necrosis factor, *PI* propidium iodide, *LDH* lactate dehydrogenase

to be released by neutrophils during secondary necrosis, but not in response to microbial stimulators [12, 13]. Cell death is commonly divided into two main forms, termed apoptotic and necrotic death modes. Apoptosis is a programmed non-lytic mode of cell death that is tightly regulated through extrinsic and intrinsic major signaling pathways mediated by caspases [14], and apoptotic cells are regularly cleared by phagocytosis without triggering inflammation [15–17].

Here, we describe techniques for inducing and measuring different forms of cell death in cells of the monocyte/macrophage lineage. In particular, we cover techniques for examining apoptosis, necrosis, pyroptosis, and necroptosis (Table 1).

Early signs of apoptosis include loss of cell membrane asymmetry, and a common method to detect this is through the use of dyes such as Annexin V to detect phosphatidylserine (PS) in the membrane, which during apoptosis become externalized and provide both a cellular signal for phagocytosis and a detectable signal for monitoring apoptosis via flow cytometry [18, 19]. In contrast to apoptosis, necrosis is largely non-programmed and is induced by external factors such as toxic chemicals, tissue damage, or disease. Many studies have reported that by using different doses of cytotoxic anticancer drugs or inducing cell stress (for example, heat, radiation, hypoxia), both apoptotic and necrotic death types could be induced simultaneously in an individual cell population [14, 20, 21]. This can also be detected via flow cytometry methods outlined in this chapter where the addition of cell viability reagents including propidium iodide (PI) to Annexin V staining can identify apoptotic cells (Annexin V positive and PI negative cells) from necrotic cells (Annexin V and PI double-positive cells), where PI can penetrate the plasma membrane.

Pyroptosis and necroptosis are forms of non-apoptotic cell death, where the cell swells and membrane rupture releases cellular contents [22]. Pyroptosis is a form of programmed necrosis which is caspase-dependent and can be activated by microbial pathogens [23]. Detecting the caspase-1 dependency using caspase-1 inhibitor can be used to distinguish pyroptosis from other necrotic cell death forms as outlined in Subheading 3.3, and additionally so is the measurement of IL-1β via ELISA [24–29].

Necroptosis is initiated through several internal and external ligand–receptor interactions. Similar to pyroptosis, it can result in organ swelling, membrane rupture, and release of cell contents. For necroptosis to occur, kinase activity of receptor-interacting protein 1 (RIP1) is required and is induced by TNF-α by binding to its receptor (TNFR1), which causes the recruitment of the TNF-receptor-associated death domain (TRADD), RIPK1, and ubiquitin E3 ligases to form a complex [30]. It is then the deubiquitination of RIPK1 that causes its disassociation from this complex and a formation of a new complex, termed complexIIb (the necrosome). Caspase-8 then needs to be inhibited for necroptosis to occur. Necroptosis can be experimentally induced by endogenous protein called second mitochondria-derived activator of caspases (SMAC) or artificial SMAC mimetics in combination with caspase inhibitors. Necroptosis can be inhibited by suppression of RIP1 kinase activity with necrostatin-1, which can be done experimentally as outlined in Subheading 3.4 [31–34]. The detection of pyroptosis and necroptosis is via the lactate dehydrogenase (LDH) assay. This assay measures the release of this enzyme upon cell death due to plasma membrane damage, with LDH activity proportional to cell lysis.

This chapter outlines the measurement of various forms of cell death and how measuring the level of MIF is a tool which can be correlated with necrotic cell death regardless of the necrosis form and involved pathways. Blocking necrosis is also associated with suppression of MIF release. In summary, measuring MIF release/level can be considered as a biomarker of necrotic cell death and tissue injury and is a beneficial additional measurement to be performed in cell death assays.

2   Materials

2.1 Cells and Medium

1. Human monocyte THP-1 leukemic cells (ATCC TIB-202, *see* **Note 1**).
2. Complete medium: RPMI 1640 media supplemented with 10% heat-inactivated fetal bovine serum (FBS), 100 U/mL penicillin, 100 μg/mL streptomycin, and 2 mM L-glutamine.
3. Tissue culture flasks.
4. Tissue culture plates (6 well and 96 well).

2.2 Inducing Apoptosis with Staurosporine

1. Staurosporine (1 μM final concentration).
2. Pan caspase inhibitor, Z-VAD-FMK (50 μM final concentration).

2.3 Inducing Necrosis with Ethanol

1. Ethanol 10% (v/v, final concentration).
2. Pan caspase inhibitor, Z-VAD-FMK (50 μM final concentration).

2.4 Inducing Pyroptosis with Nigericin

1. Nigericin (10 μM final concentration).
2. Lipopolysaccharide (100 ng/mL final concentration).
3. Caspase-1 inhibitor, Z-YVAD-FMK (25 μM final concentration).

2.5 Inducing Necroptosis

1. SMAC-mimetic BV-6 (IAP antagonist, 1 μM final concentration, *see* **Note 2**).
2. Necrostatin-1 (30 μM final concentration, *see* **Note 3**).
3. Recombinant human TNF-α (30 ng/mL final concentration).

2.6 Flow Cytometry Antibodies and Reagents

1. Annexin V-FITC (100 μg/mL stock concentration).
2. Propidium iodide (1 mg/mL stock concentration).
3. Counting particles (5.3 μm, *see* **Note 4**).
4. Phosphate-buffered saline (PBS): 137 mM NaCl, 10 mM Na_2HPO_4, 1.8 mM KH_2PO_4, and 2.7 mM KCl in milliQ water, adjust to pH 7.4 with HCl. Sterilize by autoclaving.

5. FACS buffer: PBS + 2% (v:v) FBS (filter sterilized). Store at 4 °C.

6. Binding buffer: 0.1 M Hepes (pH 7.4), 1.4 M NaCl, and 25 mM $CaCl_2$ in PBS. Sterile filter with a 0.2 µm syringe tip filter.

7. FACS tubes: 5 mL Polypropylene flow cytometry tubes capped.

2.7 Colorimetric-Based Lactate Dehydrogenase (LDH) Release Assay

1. LDH assay kit (*see* **Note 5**).

3   Methods

All procedures are carried out at room temperature unless otherwise specified. All procedures should be carried out in a class II biological hazard cabinet to maintain a sterile environment unless otherwise specified.

3.1 Inducing Apoptosis with Staurosporine

1. Seed THP-1 cells (1×10^5 cells/well) in round bottom 96-well plate in 160 µL in complete media (*see* **Note 6**).

2. Keep a set of triplicate wells as media-alone control wells for the assay. These wells complete the same experimental procedure outlined below with the exception that media is added to the wells at each step instead of reagents (*see* **Note 7**).

3. To one set of triplicate wells, add Z-VAD-FMK (50 µM final concentration) to create a set of control wells with pan caspase inhibitor. To the other wells add the same volume of media. Incubate at 37 °C in CO_2 incubator for 30 min.

4. Add staurosporine (1 µM final concentration) to triplicate wells (*see* **Note 8**). Incubate for 1–6 h at 37 °C in a CO_2 incubator.

5. Centrifuge the cells at $400 \times g$ for 5 min. Harvest the cell pellets for flow cytometry-based cell death assay (Subheading 3.5). If required, supernatants can be collected and stored at $-20/-80$ °C for analysis of secreted factors (e.g., by ELISA).

3.2 Inducing Necrosis with Ethanol

1. Seed THP-1 cells (1×10^5 cells/well) in round bottom 96-well plate in 160 µL in complete media.

2. Keep a set of triplicate wells as media-alone control wells for the assay. These wells complete the same experimental procedure outlined below with the exception that media is added to the wells at each step instead of reagents (*see* **Note 7**).

3. To one set of triplicate wells, add Z-VAD-FMK (50 µM final concentration) to create a set of control wells with pan caspase inhibitor. To the other wells add the same volume of media. Incubate at 37 °C in CO_2 incubator for 30 min.

4. Add ethanol 10% (v/v) in triplicate wells. Incubate for 1–6 h at 37 °C in a CO_2 incubator.

5. Centrifuge the cells at 400 × g for 5 min. Harvest the cell pellets for flow cytometry-based cell death assay (Subheading 3.5). If required, supernatants can be collected and stored at $-20/-80$ °C for analysis of secreted factors (e.g., by ELISA).

3.3 Inducing Pyroptosis with Nigericin

1. Seed THP-1 cells (1×10^5 cells/well) in flat bottom 96-well plate in 160 µL in complete media.

2. Keep a set of triplicate wells as media-alone control wells for the assay. These wells complete the same experimental procedure outlined below with the exception that media is added to the wells at each step instead of reagents (*see* **Note 7**).

3. Add LPS (100 ng/mL final concentration) in a total volume of 200 µL to triplicate wells and incubate overnight (16–18 h) at 37 °C in a CO_2 incubator. If required, keep a set of triplicate wells for LPS-only treated wells, to which media is added instead in steps below.

4. Centrifuge plate at 400 × g for 5 min and remove the media.

5. Add 180 µL of media to all wells to be treated with a caspase inhibitor. To these wells, add 20 µL Z-YVAD-FMK (25 µM final concentration). To all other wells add 200 µL of media. Incubate for 30 min at 37 °C in a CO_2 incubator.

6. Add nigericin (10 µM final concentration) to LPS-stimulated wells in triplicate and create a time course with hourly intervals for 1–3 h at 37 °C in a CO_2 incubator. Create a separate plate for each time point.

7. At each hour time point, centrifuge the cells at 400 × g for 5 min and collect supernatant for cytokine assays (*see* **Note 9**) and measurement of cell death by LDH assay (Subheading 3.6). Supernatants can be stored at $-20/-80$ °C until assays are performed.

3.4 Inducing Necroptosis

1. Seed THP-1 cells (5×10^5 cells/well) in round bottom 96-well plate in 160 µL in complete media.

2. Keep a set of triplicate wells as media-alone control wells for the assay. These wells complete the same experimental procedure outlined below with the exception that media is added to the wells at each step instead of reagents (*see* **Note 7**).

3. Add necrostatin-1 (30 μM final concentration) in triplicate wells to create a necroptosis inhibitor control group. To all other wells add the same volume of media.

4. To all wells add a combination of recombinant human TNF-α (30 ng/mL final concentration) and SMAC-mimetic BV-6 (IAP antagonist, 1 μM final concentration) for 18 h.

5. Centrifuge the cells at 400 × g for 5 min and collect supernatant for LDH release assay (Subheading 3.6). Supernatants can be stored at −20/−80 °C until assays are performed. If required, supernatants can also be used for analysis of secreted factors (e.g., by ELISA).

3.5 Annexin V and PI Dual Staining for Analysis of Apoptosis and Necrosis by Flow Cytometry

1. Following stimulation as outlined in the above methods for inducing apoptosis with staurosporine (Subheading 3.1) and inducing necrosis with ethanol treatment (Subheading 3.2), harvest the cells from each well and move them to a 96-well v-bottom plate.

2. Add 10 μL counting particles (*see* **Notes 3** and **10**) to 100 μL of cells.

3. Centrifuge plate at 400 × g for 5 min.

4. Wash cells once in 100 μL sterile PBS.

5. Centrifuge plate at 400 × g for 5 min.

6. Remove the supernatant, then resuspend cells in 100 μL binding buffer.

7. Add 5 μL of Annexin V FITC (100 μg/mL stock concentration) to 100 μL of the cell suspension.

8. Incubate for 15 min at room temperature, protected from light (covered in foil).

9. Add 1 μL of propidium iodide staining solution (stock concentration 1 mg/mL, final concentration 10 μg/mL) to each sample (*see* **Note 11**).

10. Incubate for 5 min at room temperature, protected from light (covered in foil).

11. Analyze by flow cytometry straight away using an instrument able to detect FITC and PI collecting 10,000 events per sample.

12. Analyze flow cytometric data with flow cytometry analysis software. Cells positive for Annexin V and negative for PI indicate early apoptosis. Double-positive cells indicate late apoptosis/necrotic stages of cell death.

The absolute cell count is then determined with the following equation:

CALCULATION:

$$(A/B) \times (C/D) = \text{Number of cells per } \mu L$$

where: A = number of events for the test sample.

B = number of events for the particles.

C = number of particles per 10 μL.

D = volume of test sample initially used in μL (100 μL as per above protocol).

3.6 Colorimetric-Based Lactate Dehydrogenase (LDH) Release Assay

1. Carry out experiments for inducing pyroptosis (Subheading 3.3) and necroptosis (Subheading 3.4) as per methods above. In addition, for the LDH assay include the following controls: a no-cell control (this will determine background readings of the culture media), an untreated cell control (*see* **Note 12**), and a maximum LDH release control (add 10 μL of 10× lysis solution per 100 μL of untreated cells, 45 min prior to adding CytoTox 96R reagent) to allow calculation of % cytotoxicity.

2. 45 min prior to harvesting supernatants (from the pyroptosis and necroptosis experiments, Subheadings 3.3 and 3.4), induce cell lysis in the set of maximum LDH release control wells by adding 10 μL of lysis solution (10×) for every 100 μL of cells.

3. Centrifuge plate at $400 \times g$ for 5 min.

4. Collect 100 μL of the supernatant for LDH release assay.

5. Transfer 50 μl of the supernatant from all wells using a mulitch-annel pipetter to a fresh 96-well flat-bottom (enzymatic assay) plate to perform the LDH release assay. The remaining 50 μL of supernatent can be stored at −80 °C until required.

6. Add 50 μL of CytoTox 96R Reagent to each well of the enzymatic assay plate containing the samples.

7. Cover the plate with foil to protect it from light and incubate for 30 min at room temperature.

8. Add 50 μL of stop solution to each well.

9. If any large bubbles are present, pop these with sterile tip.

10. Read the absorbance at 490 nm, using a microplate reader, within 1 h of adding the stop solution.

11. Results are presented as cytotoxicity percentage. For calculation of results use the following formula.

    First subtract the averaged absorbance values of the cell culture medium control wells (no-cell wells) from all absorbance values of all wells.

    CALCULATION:

$$\%\text{Cytotoxicity} = (A - B/C - B) \times 100$$

where: A = Average absorbance of experimental wells;

B = Average absorbance of untreated control wells;

C = Average absorbance of maximum LDH release control wells;

4   Notes

1. Beware of cell line variability for the induction of different forms of cell death. Some cell lines may have varying caspases and some may lack function. Perform an initial dose response curve to ensure that appropriate concentration of compound for your cell line is used.

2. SMAC-mimetic BV-6 (IAP antagonist) reduces polyubiquitination of RIP1 and increases sensitivity to cell death.

3. Necrostatin-1 is a specific inhibitor of RIP1 kinase-inducing necroptosis.

4. Counting beads for use in flow cytometry are available from a range of scientific companies for purchase.

5. LDH assay kits are commercially available to quantify cell viability via the release of LDH. It is an enzymatic assay allowing the quantification of released LDH to be measured via a colorimetric assay. The amount of color formed is proportional to the number of lysed cells.

6. To perform these assays, THP-1 cells can be seeded at a range of $1\text{--}5 \times 10^5$ cells/well, keeping cell number consistent between wells in a single experiment.

7. The untreated control cell wells may require any solvent/vehicle used (for the delivery of any reagents/test compounds) added to the well at the same concentration, for example, DMSO.

8. Prior to performing staurosporine experiments optimize the concentration on your cells and cell number of choice.

9. For pyroptosis assays, release of IL-1β can be used as a readout for inflammasome activation.

10. The use of counting beads can assist the standardization between cell death assays and determination of absolute cell count via the equation in Subheading 3.5.

11. PI staining alone, without Annexin V staining, can be performed to indicate late apoptosis/necrosis. PI is a membrane-impermeant dye which is largely excluded from viable cells and binds to double-stranded DNA.

12. The untreated control cell well for the LDH assay should have any solvent used (for the delivery of any reagents/test compounds) added to the well at the same concentration, for example, DMSO.

References

1. Calandra T, Roger T (2003) Macrophage migration inhibitory factor: a regulator of innate immunity. Nat Rev Immunol 3:791

2. Bloom J, Sun S, Al-Abed Y (2016) MIF, a controversial cytokine: a review of structural features, challenges, and opportunities for drug development. Expert Opin Ther Targets 20(12):1463–1475

3. Calandra T et al (1994) The macrophage is an important and previously unrecognized source of macrophage migration inhibitory factor. J Exp Med 179(6):1895–1902

4. Rice EK et al (2003) Induction of MIF synthesis and secretion by tubular epithelial cells: a novel action of angiotensin II. Kidney Int 63 (4):1265–1275

5. Calandra T et al (2000) Protection from septic shock by neutralization of macrophage migration inhibitory factor. Nat Med 6:164

6. Radstake TRDJ et al (2005) Correlation of rheumatoid arthritis severity with the genetic functional variants and circulating levels of macrophage migration inhibitory factor. Arthritis Rheum 52(10):3020–3029

7. Kim HR et al (2007) Macrophage migration inhibitory factor upregulates angiogenic factors and correlates with clinical measures in rheumatoid arthritis. J Rheumatol 34 (5):927–936

8. Foote A et al (2004) Macrophage migration inhibitory factor in systemic lupus erythematosus. J Rheumatol 31(2):268–273

9. Sánchez E et al (2006) Evidence of association of macrophage migration inhibitory factor gene polymorphisms with systemic lupus erythematosus. Genes Immun 7:433

10. Kok T et al (2018) Small-molecule inhibitors of macrophage migration inhibitory factor (MIF) as an emerging class of therapeutics for immune disorders. Drug Discov Today 23 (11):1910–1918

11. Hertelendy J et al (2018) Macrophage migration inhibitory factor - a favorable marker in inflammatory diseases? Curr Med Chem 25 (5):601–605

12. Roth S et al (2015) Secondary necrotic neutrophils release interleukin-16C and macrophage migration inhibitory factor from stores in the cytosol. Cell Death Discov 1:15056–15056

13. Roth S, Solbach W, Laskay T (2016) IL-16 and MIF: messengers beyond neutrophil cell death. Cell Death Dis 7(1):e2049–e2049

14. Elmore S (2007) Apoptosis: a review of programmed cell death. Toxicol Pathol 35 (4):495–516

15. Jorgensen I, Rayamajhi M, Miao EA (2017) Programmed cell death as a defence against infection. Nat Rev Immunol 17:151

16. Cohen JJ (1991) Programmed cell death in the immune system. Adv Immunol 50:55–85

17. Yang Y et al (2015) Programmed cell death and its role in inflammation. Mil Med Res 2(1):12

18. Choi JJ, Reich Iii CF, Pisetsky DS (2004) Release of DNA from dead and dying lymphocyte and monocyte cell lines in vitro. Scand J Immunol 60(1–2):159–166

19. Vermes I et al (1995) A novel assay for apoptosis. Flow cytometric detection of phosphatidylserine expression on early apoptotic cells using fluorescein labelled Annexin V. J Immunol Methods 184(1):39–51

20. Yurinskaya V et al (2017) A comparative study of U937 cell size changes during apoptosis initiation by flow cytometry, light scattering, water assay and electronic sizing. Apoptosis 22(10):1287–1295

21. Castilla R et al (2004) Dual effect of ethanol on cell death in primary culture of human and rat hepatocytes. Alcohol Alcohol 39(4):290–296

22. Tait SWG, Ichim G, Green DR (2014) Die another way--non-apoptotic mechanisms of cell death. J Cell Sci 127(Pt 10):2135–2144

23. Bergsbaken T, Fink SL, Cookson BT (2009) Pyroptosis: host cell death and inflammation. Nat Rev Microbiol 7(2):99–109

24. Man SM, Karki R, Kanneganti TD (2017) Molecular mechanisms and functions of pyroptosis, inflammatory caspases and inflammasomes in infectious diseases. Immunol Rev 277(1):61–75

25. Magna M, Pisetsky D (2015) The role of cell death in the pathogenesis of SLE: Is pyroptosis the missing link? Scand J Immunol 82 (3):218–224

26. Aglietti RA, Dueber EC (2017) Recent insights into the molecular mechanisms underlying pyroptosis and gasdermin family functions. Trends Immunol 38(4):261–271

27. Vanaja SK, Rathinam VA, Fitzgerald KA (2015) Mechanisms of inflammasome activation: recent advances and novel insights. Trends Cell Biol 25(5):308–315

28. He Y, Hara H, Nunez G (2016) Mechanism and regulation of NLRP3 inflammasome activation. Trends Biochem Sci 41 (12):1012–1021

29. Lang T et al (2018) Macrophage migration inhibitory factor is required for NLRP3 inflammasome activation. Nat Commun 9(1):2223

30. Dhuriya YK, Sharma D (2018) Necroptosis: a regulated inflammatory mode of cell death. J Neuroinflammation 15(1):199

31. Omoto S et al (2015) Suppression of RIP3-dependent necroptosis by human cytomegalovirus. J Biol Chem 290(18):11635–11648

32. Murphy JM, Silke J (2014) Ars Moriendi; the art of dying well–new insights into the molecular pathways of necroptotic cell death. EMBO Rep 15(2):155–164

33. Hanson B (2016) Necroptosis: a new way of dying? Cancer Biol Ther 17(9):899–910

34. Kearney CJ, Martin SJ (2017) An inflammatory perspective on necroptosis. Mol Cell 65 (6):965–973

Chapter 16

Assessing the Role of MIF in *Plasmodium* spp. Infections Using Ex Vivo Models

Elizabeth H. Aitken

Abstract

Ex vivo techniques are a valuable tool for the investigation of how immune cells respond to *Plasmodium* spp. antigen, allowing examination of various aspects of the immune response under controlled conditions. Here we describe how to isolate peripheral blood mononuclear cells (PBMC) from donors and coculture them with purified *P. falciparum*-infected red blood cells (iRBC) to investigate the role of MIF during *Plasmodium* spp. infection.

Key words *Plasmodium falciparum*, Peripheral blood mononuclear cell (PBMC), Malaria, Cytokine

1 Introduction

Ex vivo experiments have been used to help untangle various aspects of host cell responses during *Plasmodium* spp. infection, including recently the role of macrophage migration inhibitory factor (MIF) in peripheral blood mononuclear cells (PBMC) Interleukin 1β secretion in response to *P. falciparum*-infected red blood cells (iRBC) [1]. Ex vivo techniques are a valuable tool for the investigation of how immune cells respond to *Plasmodium* spp. antigen. Not only do you have control over experimental conditions, but you can also target cells or pathways of interest (for example, by using agonists, antagonists, or inhibitors). Because multiple wells can be set up, there is little limit in the number of outputs of interest that can be measured. Finally, and importantly, ex vivo techniques use human tissues which can help validate findings from murine models. This chapter will describe methods used to collect and isolate human PBMC and prepare *P. falciparum* antigen for subsequent stimulation of PBMC to allow the investigation of the role of MIF during *Plasmodium* spp. infection.

James Harris and Eric F. Morand (eds.), *Macrophage Migration Inhibitory Factor: Methods and Protocols*,
Methods in Molecular Biology, vol. 2080, https://doi.org/10.1007/978-1-4939-9936-1_16,
© Springer Science+Business Media, LLC, part of Springer Nature 2020

2　Materials

All plasticware must be sterile and all reagents need to be tissue culture grade and kept sterile.

2.1　Equipment

1. Benchtop centrifuge.
2. Class II biological safety cabinet.
3. $-80\ ^\circ$C Freezer.
4. Water bath at 37 $^\circ$C.
5. Light microscope with $1000\times$ magnification.
6. Inverted light microscope with $400\times$ magnification.
7. 37 $^\circ$C, 5% CO_2 incubator.
8. Liquid nitrogen storage.
9. Hemocytometer.
10. Thermo Scientific "Mr. Frosty" freezing container.
11. Pipette gun with 10 and 25 mL pipettes.
12. P200 Micropipette and 200 µL pipette tips.
13. 3 mL Transfer pipettes.
14. Glass microscope slides.
15. Tube racks for 15 mL, 50 mL, and cryotubes.
16. 15 mL and 50 mL conical falcon tubes and 1.5 mL cryotubes.
17. 96-well round-bottom plates with lids.
18. Blood collection equipment: BD lithium heparin 6 mL vacutainers, BD Vacutainer blood collection needle 22G, and holder.

2.2　Reagents

1. Density gradient solution (density 1.077 g/mL). For example, GE Healthcare Ficoll-Paque Plus.
2. Phosphate-buffered saline solution (PBS; $1\times$ and $100\times$): Without Mg^+ and Ca^+.
3. Trypan blue solution: 0.4% Trypan blue in PBS. Filtered and stored at room temperature.
4. Heat-inactivated fetal calf serum (HI FCS) (*see* **Note 1**). Aliquoted and stored at $-20\ ^\circ$C.
5. Freezing solution: 20% Dimethyl sulfoxide (DMSO) and 80% cold HI FCS. Filtered before use.
6. Liquid nitrogen.
7. Microscope oil.
8. RPMI-1640 10% FCS media: 450 mL of RPMI-1640 with phenol red, 50 mL HI FCS, 5 mL Gibco penicillin-

streptomycin-glutamine (100×). Store at 4 °C and warm to 37 °C before use.

9. RPMI-1640 with HEPES: 500 mL RPMI-1640 with HEPES buffer (25 mM). Store at 4 °C and warm to 37 °C before use.

10. Percoll gradient solutions: Percoll gradients are made up by mixing 90% Percoll solution, RPMI-1640 with HEPES, and RPMI-1640 with HEPES and sorbitol. To make 90% Percoll solution, mix 180 mL GE Healthcare Percoll and 20 mL PBS (10×). To make RPMI-1640 with HEPES and sorbitol, mix 12 g sorbitol with 33 mL RPMI-1640 with HEPES at 37 °C until dissolved, filter. For 80% Percoll, mix 89 mL of 90% Percoll with 11 mL of RPMI-1640 with HEPES and sorbitol. For 60% Percoll, mix 67 mL 90% Percoll, 11 mL of RPMI-1640 with HEPES and sorbitol, and 22 mL RPMI-1640 with HEPES. For 40% Percoll, mix 44 mL 90% Percoll, 11 mL of RPMI-1640 with HEPES and sorbitol, and 45 mL RPMI-1640 with HEPES. Store all solutions at 4 °C and warm to 37 °C before use.

11. *P. falciparum* parasite culture. Parasites can be cultured in vitro in RPMI-1640, supplemented with 25 mM HEPES, 0.2%w/vol $NaHCO_3$, 5% pooled heat-inactivated non-immune human sera, and 0.25% Albumax. Parasites should be cultured in group O red blood cells and in a low oxygen environment (using a low gas mixture) at 37 °C. For detailed procedures on parasite culture please see papers by Jensen et al. [2, 3].

12. Methanol: Laboratory reagent grade.

13. Giemsa stain: 10 mL Merck Giemsa's azur eosin methylene blue solution, 90 mL water. Make fresh before use.

14. Glycerolyte: 57 mL Glycerol, 1.6 g sodium lactate ($C_3H_5NaO_3$), 30 mg potassium chloride (KCl), 1.38 g sodium dihydrogen phosphate (NaH_2PO_4), top up to 100 mL with ddH_2O. Adjust pH to 6.8. Store at 4 °C.

15. Parasite thawing solutions: Thawing solution A, PBS with an extra 2.8 g of NaCl per 100 mL. Thawing solution B, a mix of 50% PBS and 50% thawing solution A.

---

3   Methods

3.1   PBMC Collection and Storage

1. Collect whole blood in a 6 mL-tube with anticoagulant, such as heparin.

2. Using the pipette gun and 10 mL pipette, transfer whole blood into a 50 mL conical tube.

3. Dilute blood with an equal volume of PBS.

4. In a new 50 mL conical tube, aliquot 15 mL of Ficoll-Paque.

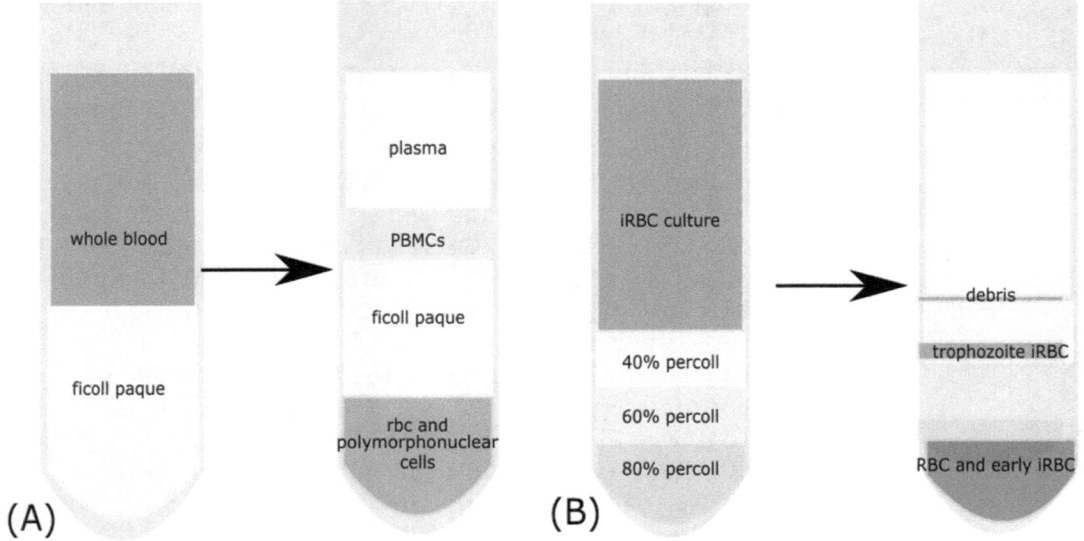

Fig. 1 Diagram with layers labeled before and after centrifugation for (**a**) peripheral blood mononuclear cell (PBMC) isolation and (**b**) infected red blood cell (iRBC) purification

5. Carefully layer diluted blood on top of Ficoll-Paque using a transfer pipette (Fig. 1a).

6. Centrifuge at $800 \times g$ for 20 min with the break off.

7. Remove the PBMC layer (between the plasma and Ficoll-Paque layers (Fig. 1a) into a new 50 mL conical tube using a transfer pipette.

8. Top up 50 mL tube containing PBMC with PBS.

9. Centrifuge at $250 \times g$ for 10 min.

10. Aspirate supernatant, resuspend cells in 40 mL of PBS.

11. Take 100 µL aliquot and use it to count the number of live cells using a hemocytometer and trypan blue exclusion on the inverted light microscope with ×400 magnification.

12. Centrifuge 50 mL tube with PBMC in PBS at $250 \times g$ for 10 min.

13. Remove supernatant and resuspend at 2×10^7 cells/mL in 100% cold HI FCS.

14. Add an equal volume of freezing solution dropwise to get a final concentration of 1×10^7 cells/mL.

15. Aliquot into cryovials and freeze at $-80\ °C$ in a Mr. Frosty for 24 h.

16. Put into liquid nitrogen for long-term storage.

3.2 P. falciparum Preparation and Purification

P. falciparum blood stage parasites can be efficiently purified directly from in vitro culture and if necessary stored as frozen stock until use (*see* **Note 2**). This chapter describes the purification of cultured *P. falciparum* parasites for ex vivo stimulation of PBMC. However, before use parasite cultures should be checked for mycoplasma contamination (*see* **Note 3**), and, if relevant, the binding phenotype of the iRBC should also be assessed (*see* **Note 4**).

1. In a 15 mL tube carefully layer 2 mL each of 80%, 60%, and 40% Percoll solution on top of each other (*see* **Note 5**) (Fig. 1b).

2. Centrifuge parasite culture at $500 \times g$ for 5 min; for each 25 mL of culture resuspend in 2 mL of RPMI-1640 with HEPES.

3. Layer parasite mixture on top of Percoll gradients (Fig. 1b).

4. Centrifuge for $1600 \times g$ for 15 min.

5. Any merozoites will sit at the top of the gradient, trophozoites in the 60% layer and uninfected cells at the bottom (Fig. 1b). If there are two bands in the 60% layer, they are usually mature trophozoites and immature trophozoites (in the top and bottom bands, respectively).

6. Collect the layer containing trophozoites.

7. Wash three times with RPMI-1640 with HEPES by centrifuging at $500 \times g$ for 5 min and discarding supernatant. Count the number of iRBC isolated using a hemocytometer on the inverted light microscope with $\times 400$ magnification.

8. To determine parasitemia of purified iRBC, prepare a thin smear on a glass slide, dry, fix with methanol for 1 min, dry and stain with Giemsa for 3 min and then rinse with running water. Observe dried slide using light microscopy $\times 1000$ with microscope oil. Usually a parasitemia well above 80% can be obtained using this method.

9. Purified parasites can be used immediately or frozen down for later use (*see* **Note 2**). To freeze, pellet parasites by centrifugation at $500 \times g$ for 5 min, discard supernatant, then resuspend parasite pellet dropwise so the final concentration is 1 part parasite pellet to 2 parts glycerolyte. Aliquot glycerolyte:parasite suspension into cryovials and freeze at -80 °C in a Mr. Frosty for 24 h. Put into liquid nitrogen for long-term storage.

10. To thaw frozen purified parasites, take vial from liquid nitrogen and thaw at room temperature. Aliquot parasite solution from cryovial into a 15 mL tube, noting the volume. Using a P200 pipette slowly add drop by drop an equal volume of thawing

solution A, gently shaking the tube to resuspend the solution. Leave for 2 min and then add drop by drop an additional 2 mL of thawing solution A. Centrifuge at $500 \times g$ for 5 min, discard supernatant, and then slowly add 2 mL of thawing solution B shaking gently to resuspend. Centrifuge at $500 \times g$ for 5 min and discard supernatant and then add 2 mL of PBS drop by drop. Centrifuge at $500 \times g$ for 5 min and discard supernatant and resuspend thawed parasites in PBS.

3.3 Parasite and PBMC Coculture

1. Thaw vial of frozen PBMC quickly in water bath at 37 °C.

2. Immediately aliquot into a 15 mL conical flask and top up to 10 mL with RPMI-1640.

3. Pellet PBMC by centrifugation at $250 \times g$ for 10 min.

4. Resuspend in 10 mL RPMI-1640, take 100 µL aliquot and use it to count the number of live cells using a hemocytometer and trypan blue exclusion on the Inverted light microscope under $\times 400$ magnification.

5. Re-pellet remaining PBMC by centrifugation at $250 \times g$ for 10 min.

6. Resuspend PBMC at 2×10^6 viable PBMC/mL in RPMI-1640 10% FCS media.

7. Plate out PBMC 100 µL/well (2×10^5 viable PBMC/well) in a 96-well round-bottom plate (*see* **Note 6**).

8. Resuspend purified trophozoite stage iRBC (and non-infected RBC controls) at a concentration of 2×10^7 cells/mL in RPMI-1640 10% FCS media.

9. Add 100 µL resuspended iRBC or RBC solution to PBMC (2×10^6 RBC/well).

10. To investigate the role of MIF add desired MIF inhibitor, for example, ISO-1 can be used to inhibit MIF in parasite PBMC cocultures at concentrations of 5–50 µM [1].

11. Run each condition in triplicate and with multiple PBMC donors, culture plate at 37 °C 5% CO_2.

12. Depending on output of interest, stimulation of PBMC should be halted at different time points. For cytokine secretion 16 h is sufficient though time points up to 48 h can be used.

13. To stop stimulation spin plate at $250 \times g$ for 5 min at 4 °C.

14. Supernatants can then be collected and analyzed for cytokine secretion (*see* **Notes 7** and **8**).

4 Notes

1. FCS quality varies between suppliers and even between batches, and in our experience certain batches have even been known to activate white blood cells. Ideally experiments should be done with a single batch of FCS, and before purchasing a new batch of FCS it should be tried out in your own lab culture systems.

2. There is often significant lysis and loss of purified parasites during the thawing procedure; however if you do not have a parasite culture system set up and are relying on purified parasites from another source, then this may still be a viable option.

3. Testing *P. falciparum* culture for mycoplasma prior to using it to investigate cellular immune responses to parasite antigen is necessary. In our experience PCR techniques were not sufficient to identify some contaminations. Commercial kits which test for mycoplasma enzymes such as Lonza's MycoAlert are efficient and staining for mycoplasma using Hoerchst is another option.

4. *P. falciparum* can express a variety of antigens on the surface of the iRBC, which in turn can alter their binding phenotype. This is important because some binding phenotypes are associated with specific diseases, for example, chondroitin sulfate-A binding is associated with placental malaria [4, 5] and endothelial protein C receptor binding is associated with severe malaria [6]. It is also important because parasites can also express surface antigens which allow them to bind CD36 which is also expressed on monocytes [7]. If you desire cultured *P. falciparum* to have a specific phenotype for your experiments you can select the desired phenotype by FACS sorting using a tagged monoclonal to the antigen of interest, pan parasites on purified receptors coated on plates or expressed on cells or use parasite lines which have mutations resulting in the expression of the desired phenotype.

5. Layer Percoll only immediately before use, otherwise layers will not be distinct after centrifugation.

6. If possible do not use outside wells for cell culture, instead use inside wells and fill outside wells with 200 μL of sterile PBS.

7. As well as purified whole iRBC other parasite antigen sources can be used to stimulate PBMC, for example, schizont extract (obtained from freeze thawing purified schizont stage iRBC) or parasite hemozoin.

8. Cells can also be collected and examined for outcomes of interest. For example, they can be stained for intracellular cytokines, cell surface markers of interest, with trypan blue to

look at viability or put onto slides and stained with simple reagents such as Diff-Quick or Giemsa and viewed by microscopy to examine morphology and antigen uptake.

References

1. Lang T et al (2018) Macrophage migration inhibitory factor is required for NLRP3 inflammasome activation. Nat Commun 9(1):2223

2. Jensen JB (2002) In vitro culture of plasmodium parasites. In: Doolan DL (ed) Malaria methods and protocols. Humana Press, Totowa, NJ, pp 477–488

3. Trager W, Jensen JB (1976) Human malaria parasites in continuous culture. Science 193 (4254):673–675

4. Fried M, Duffy PE (1996) Adherence of Plasmodium falciparum to chondroitin sulfate A in the human placenta. Science 272 (5267):1502–1504

5. Rogerson SJ, Brown GV (1997) Chondroitin sulphate A as an adherence receptor for Plasmodium falciparum-infected erythrocytes. Parasitol Today 13(2):70–75

6. Turner L et al (2013) Severe malaria is associated with parasite binding to endothelial protein C receptor. Nature 498(7455):502–505

7. Ockenhouse CF et al (1989) Identification of a platelet membrane glycoprotein as a falciparum malaria sequestration receptor. Science 243 (4897):1469–1471

Chapter 17

The Effect of Macrophage Migration Inhibitory Factor on Intestinal Permeability: FITC-Dextran Serum Measurement and Transmission Electron Microscopy

Milica Vujičić, Sanja Despotović, Tamara Saksida, and Ivana Stojanović

Abstract

Macrophage migration inhibitory factor (MIF) is a molecule with multiple functions: from enforcing the immune system to fight bacterial infection to the regulation of insulin activity. Also, MIF is expressed by enterocytes that line the intestinal border toward the lumen, and in M cells, where it regulates phagocytosis of antigens from the lumen of the gut and their transport to Peyer's patches. Since there were no data on the role of MIF in the maintenance of the intestinal barrier, we used MIF-deficient mice bred on C57BL/6 background as a model for the investigation of intestinal permeability. The obtained results indicate that the absence of MIF increases intestinal permeability. Here we describe two methods for measuring intestinal permeability in mice: detection of orally delivered FITC-dextran in the serum and transmission electron microscopy used for visualization and measurement of cell-to-cell connections width.

Key words Macrophage migration inhibitory factor, Intestinal permeability, FITC-dextran, Transmission electron microscopy

1 Introduction

Macrophage migration inhibitory factor (MIF) is a pleiotropic cytokine produced by many different cell types—immune, epithelial, and endothelial [1]. MIF's primary function is the promotion of inflammation and microbe eradication [2]. When secreted from activated cells, MIF instructs macrophages to remain at the inflammatory site until the resolution of infection. In addition, MIF is a very potent stimulus for the production of other pro-inflammatory cytokines. Therefore, it is quite expected that mice with genetic MIF deletion ($Mif^{-/-}$) display an anti-inflammatory phenotype. Their activated immune cells secrete lower amounts of pro-inflammatory IL-1β, IFN-γ, IL-17, and IL-23, and higher

Milica Vujičić and Sanja Despotović contributed equally to this work.

James Harris and Eric F. Morand (eds.), *Macrophage Migration Inhibitory Factor: Methods and Protocols*,
Methods in Molecular Biology, vol. 2080, https://doi.org/10.1007/978-1-4939-9936-1_17,
© Springer Science+Business Media, LLC, part of Springer Nature 2020

levels of the anti-inflammatory cytokines IL-4 and IL-10 [3, 4]. MIF also exerts its pro-inflammatory actions through the opposing effect it has on the function of glucocorticoids [5].

Apart from the well-documented role in inflammation, MIF also regulates glucose metabolism [6]. First, it acts as a chaperone and enables proper activity of insulin molecule within beta cells of the pancreas. Then, MIF can potentiate glucose uptake into beta cells. Finally, it was found that MIF deficiency promotes the development of glucose intolerance and obesity in C57BL/6 mice [7–9].

MIF is also expressed in the epithelial cells of the colon and in studies by Ohkawara et al., its overexpression is correlated with the pathogenesis of colitis in mice [10, 11]. However, MIF is also produced by M cells, located above Peyer's patches, and in the isolated lymphoid follicles, whose function is to phagocytose antigens from the lumen of the gut [12]. Without MIF, this antigen sampling is highly downregulated. Epithelial cells firmly connected with tight and adhesive junctions are only one part of the intestinal barrier. The other functional part of the barrier comprises immune cells and mediators secreted by endothelial cells [13]. The intestinal barrier plays a decisive role in maintenance of the homeostasis between microbiota in the lumen of the gut and the rest of the organism. Our previous study showed the involvement of MIF in the maintenance of gut microbiota diversity and immune surveillance, which in turn enables proper function of the intestinal barrier [14]. We have used $Mif^{-/-}$ mice as a model system for evaluation of intestinal barrier integrity. In the absence of MIF, intestinal permeability was increased, as indicated by the higher serum levels of orally delivered FITC-dextran. Also, the width of tight and adherens junctions between the epithelial cells of the colon was increased in $Mif^{-/-}$ mice. This was probably a result of the observed disturbance of E-cadherin, zonula occludens-1, occludin, and claudin-2 expression, all molecules involved in the maintenance of proper cell junction architecture [14].

In this chapter, we describe two methods for measurement of intestinal permeability in mice—a FITC-dextran method and transmission electron microscopy. FITC-dextran is a fluorescently labeled polymer of anhydroglucose. It is a large molecule (3–5 kDa) that is not able to cross the epithelial barrier under normal conditions [15]. However, disruption of tight junctions, as happens during intestinal inflammation, allows FITC-dextran to cross the epithelial barrier, which makes it a useful marker of intestinal permeability. This method can be difficult to perform but it gives reproducible results. The other method is based on the visualization and measurement of the width of tight and adherens junctions between the epithelial cells of the gut by transmission electron microscopy. It is a highly sensitive method that allowed us to visualize membrane and cytoskeletal components of intercellular junctions [14, 16]. It is performed on fixed tissue samples and the preservation of junctional complexes is highly dependent on the quality of fixation [16–18].

2 Materials

2.1 FITC-Dextran Method

1. FITC-dextran 4 kDa.

   For the treatment, dissolve FITC-dextran in distilled water (44 mg FITC-dextran per 100 g body weight of mouse).

   For the standard curve, make twofold dilutions of FITC-dextran in the physiological saline starting from 2000 to 0 ng/mL and use their fluorescence values to calculate a standard curve.

2. Oral gavage needle connected to a 1 mL syringe.

3. 1.5 mL conical tubes.

4. Black 96-well plate.

2.2 Transmission Electron Microscopy

1. 0.4 M Cacodylate buffer: 6.49 g of cacodylic acid-sodium salts in 100 mL distilled water.

2. 0.1 M Cacodylate buffer: 10 mL 0.4 M cacodylate buffer in 30 mL distilled water.

3. 25% Glutaraldehyde (commercial solution).

4. 2% Osmium tetroxide (OsO_4) in distilled water.

5. Fixative solution 1: 3% Glutaraldehyde in 0.1 M cacodylate buffer (pH 7.2–7.4). For 10 mL, mix 2.5 mL 0.4 M cacodylate buffer stock with 1 mL 25% glutaraldehyde and 6.5 mL distilled water.

6. Fixative solution 2: 1% OsO_4 in 0.1 M cacodylate buffer (pH 7.4). For 10 mL, mix 2.5 mL 0.4 M cacodylate buffer stock, 2.5 mL of distilled water, and 5 mL of 2% OsO_4 solution.

7. 4.7% Uranyl acetate solution (in distilled water).

8. Ethanol.

9. Propylene oxide.

10. Epoxy resins.

11. Reynolds' lead citrate solution: To prepare 50 mL of contrasting solution, mix 1.33 g of lead nitrate and 1.76 g of sodium citrate with 8 mL 1 N NaOH (diluted in distilled water) and stir for 30 min. Add 42 mL of distilled water and stir for another 15 min. The solution pH should be 12. Store in the dark for 24 h before use.

12. Toluidine Blue.

13. Ultramicrotome.

14. Grids for transmission electron microscopy.

3 Method

3.1 FITC-dextran Method for Measuring Intestinal Permeability

All steps are performed at room temperature, unless otherwise specified.

1. Put the mice on an overnight fasting. One group of mice (preferably 7–10 mice) should be treated orally with the FITC-dextran solution and the other group of maximum 3 mice should receive diluent (control group orally gavaged with distilled water). After fluorescence measurement, values from the control group should be subtracted from the values of treated group in order to annul the background fluorescence of sera.

2. Prepare FITC-dextran solution prior to use and keep it protected from light.

3. Administer 50 μL of FITC-dextran per mice by oral gavage (*see* **Note 1**) (Fig. 1).

4. Let mice rest for 4 h without food or water (*see* **Note 2**).

5. After 4 h, isolate blood from retroorbital sinus, a minimal amount of 300 μL.

6. Collect blood into the 1.5 mL conical tube and centrifuge at $20,000 \times g$ for 10 min at 4 °C (*see* **Note 3**).

7. Transfer 50 μL of each serum in triplicates into a 96-well nontransparent black plate.

8. Transfer 50 μL of FITC-dextran two-fold dilutions (previously prepared) in triplicates.

9. Measure the fluorescence of FITC-dextran on a fluorescence plate reader with the suitable wavelengths (excitation maximum 488 nm, emission maximum 534 nm). Our reader was Chameleon (Hidex, Turku, Finland).

10. Use the sera without FITC-dextran to measure the background fluorescence and subtract that value from the fluorescence intensity measured in the sera of treated mice.

11. Optional step: measure background absorbance (the redness of the serum) on absorbance wavelength of 600–700 nm (*see* **Note 4**).

12. Calculate the exact concentration of FITC-dextran in the serum according to the standard curve values obtained from FITC-dextran two-fold dilutions absorbances (Fig. 2).

3.2 Transmission Electron Microscopy (TEM) Examination and Ultrastructural Morphometry (Fig. 3)

This method consists of five steps: tissue fixation, dehydration and embedding, semi-thin and ultrathin sectioning, contrasting and morphometric measurements.

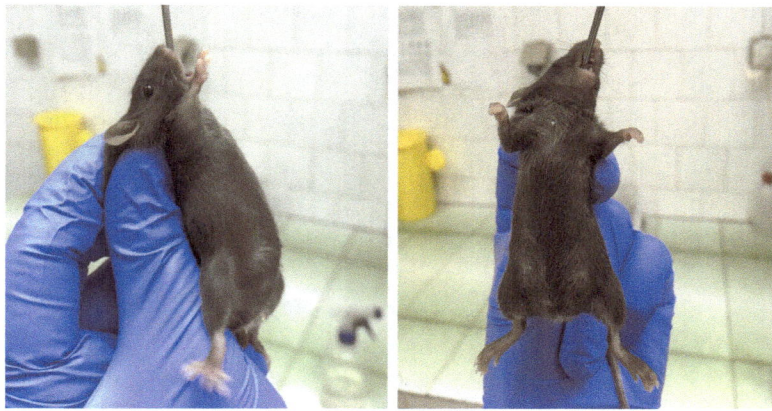

Fig. 1 Administration of FITC-dextran by oral gavage

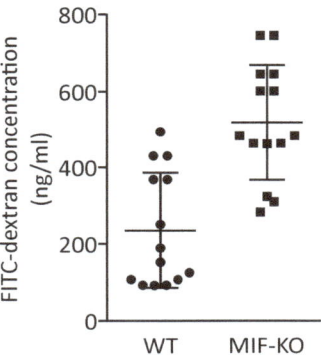

Fig. 2 Concentration of serum FITC-dextran representing intestinal permeability. The figure is reproduced from Vujicic et al., 2018, with permission from Scientific Reports (Creative Commons Attribution 4.0 International License: https:/ creativecommons.org/licenses/by/4.0/)

3.2.1 Tissue Fixation

1. The animal should be euthanized either by cervical dislocation or by CO_2 asphyxiation.

2. Cut open the animal abdominal cavity and take small samples of colon from different areas (1 cm, 2 cm and 3 cm from the rectum). Colon tissue samples have to be small in order to fix fast and completely; if the whole circumference of colon is taken, the sample length should not exceed 3–5 mm.

3. Place the colon samples immediately in the fixative solution 1 to preserve the structure of cells and cellular junctions (*see* **Note 5**).

4. Keep the samples in fixative 1 overnight on 4 °C.

5. Transfer the samples to the fixative solution 2 (*see* **Note 6**).

6. Samples should stay in 1% OsO_4 for 1–1.5 h on 4 °C.

198   Milica Vujičić et al.

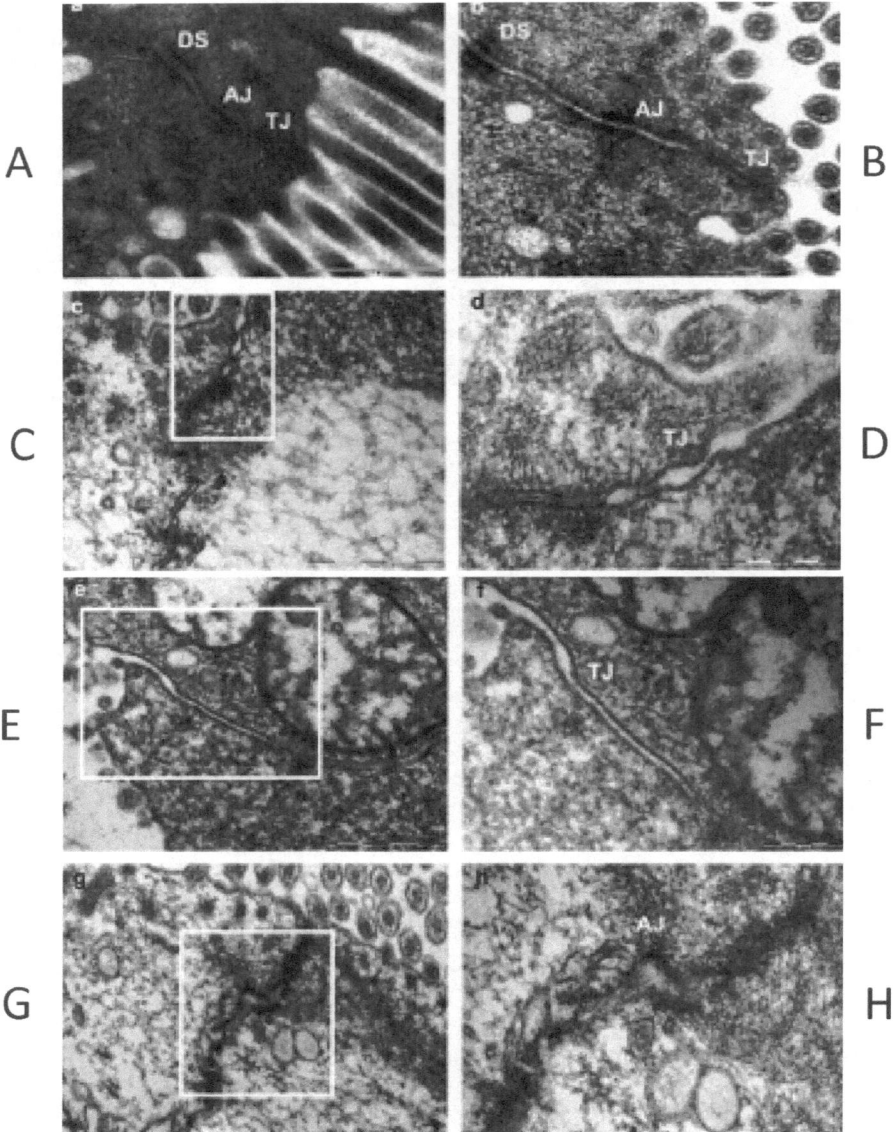

Fig. 3 The influence of MIF on the intestinal permeability. (**a–h**) Electron micrographs showing the tight junctions (TJ), adherens junctions (AJ), and desmosomes (DS) of colon epithelial cells. (**a, b**) TJ, AJ, and DS in the colon of WT mice (56,000×). (**c**) TJ in the colon of $Mif^{-/-}$ animals were wider, without complete obliteration of intercellular space (56,000×). (**d**) The same TJ in $Mif^{-/-}$ animal, on higher magnification (140,000×). (**e**) Wide TJ in the colon of $Mif^{-/-}$ animal; obliteration of intercellular space is missing (56,000×). (**f**) The same TJ in $Mif^{-/-}$ animal, on higher magnification (89,000×). (**g**) Wide AJ in the colon of $Mif^{-/-}$ animal (44,000×). (**h**) The same AJ in $Mif^{-/-}$ animal, on higher magnification (89,000×). The figure is reproduced from Vujicic et al., 2018, with permission from Scientific Reports (Creative Commons Attribution 4.0 International License: https:/creativecommons.org/licenses/by/4.0/)

7. Wash the samples (3 × 10 min) in 0.1 M cacodylate buffer.

8. Transfer the samples to 4.7% uranyl acetate solution and keep them overnight on 4 °C.

3.2.2 Dehydration and Embedding

1. Wash the colon tissue samples in distilled water for 5 mi and pass them through graded alcohols (cold 50% ethanol for 5 min, 70% ethanol for 30 min on 4 °C, 95–96% ethanol for 30 min, on 4 °C and 100% ethanol, 3 × 10 min, on room temperature) and propylene oxide (2 × 10 min).

2. Embedding in epoxy resins is gradual: transfer samples to 1:1 mixture of epoxy resins and propylene oxide, overnight on room temperature.

3. Transfer samples to 3:1 mixture of epoxy resins and propylene oxide, 3–5 h at room temperature.

4. Transfer colon tissue samples in only epoxy resins, overnight on 60 °C.

3.2.3 Semi-thin and Ultrathin Sectioning

1. Cut the resin blocks into semi-thin sections (0.5–1 μm) with a glass knife, on an ultramicrotome.

2. Stain the sections with toluidine blue for orientation on light microscopy. On semi-thin sections the small area of tissue was selected for ultrathin sectioning (*see* **Note** 7).

3. Ultrathin sections (50–70 nm) are made with a diamond knife on the ultramicrotome and collected on copper grids.

3.2.4 Contrasting

1. For contrasting, stain copper grids with 4% uranyl acetate for 25 min.

2. Rinse in distilled water four times.

3. Stain with Reynolds' lead citrate solution for 5 min and rinse again in distilled water.

3.2.5 Morphometric measurements

1. Perform ultrastructural morphometry on ultra-micrographs taken at 56,000× magnification on 50 tight and adherens junctions per tissue sample using an open-source software Fiji. The width of tight and adherens junction (in nm) between outer layers of adjacent cell membranes should be measured on two different spots along each junction. Width of adherens and tight junctions is represented by median in the interquartile range (IQR) 25th–75th percentiles (Fig. 3).

4   Notes

1. This is a crucial step because the spilling of FITC-dextran will give false-negative results. Mice should be manually restrained for the oral gavage (Fig. 1). The gavage needle (stainless steel

gavage tube or disposable flexible gavage tubes) should be inserted gently into the esophagus while avoiding excessive force.

2. FITC-dextran is stable for 24 h in vivo.

3. To ensure that the blood clot is completely separated from the serum, use a small needle or a wooden stick to gently detach the blood clot from the tube wall. This will ease the separation.

4. If erythrocytes are lysed in the process of serum preparation, they color the serum red. In this case, add another step of measuring background absorbance (the redness of the serum) on absorbance wavelength of 600–700 nm. The obtained absorbance values could be used for normalization of the fluorescence intensity of samples.

5. Colon tissue samples should be immersed in fixative solution immediately after removal from the animal. This is very important—the speed of fixation is the most critical step. If a small amount of colonic content remained in the sample, it has to be removed as soon as possible after an experiment (recommended within 2 h), by gently washing with syringe filled with fixative. In order to preserve epithelial cells, touching of mucosa should be avoided. At the same time, the sample can be remodeled, if necessary, to smaller samples, with care to be taken that each sample contains mucosa.

6. This fixative has to be made immediately before use.

7. To visualize colon epithelium, we selected the small, trapezoid-shaped part of mucosa, containing predominantly epithelial cells.

Acknowledgements

Supported by the Ministry of Education, Science and Technological Development of the Republic of Serbia (grant 173013 and 175005).

References

1. Calandra T, Roger T (2003) Macrophage migration inhibitory factor: a regulator of innate immunity. Nat Rev Immunol 3:791–800

2. Stojanovic I, Mirkov I, Kataranovski M et al (2011) A role for macrophage migration inhibitory factor in protective immunity against Aspergillus fumigatus. Immunobiology 216:1018–1027

3. Stojanovic I, Cvjeticanin T, Lazaroski S et al (2009) Macrophage migration inhibitory factor stimulates interleukin-17 expression and production in lymph node cells. Immunology 126:74–83

4. Gao XM, Liu Y, White D et al (2011) Deletion of macrophage migration inhibitory factor protects the heart from severe ischemia-reperfusion injury: a predominant role of anti-inflammation. J Mol Cell Cardiol 50:991–999

5. Flaster H, Bernhagen J, Calandra T et al (2007) The macrophage migration inhibitory factor-glucocorticoid dyad: regulation of

inflammation and immunity. Mol Endocrinol 21:1267–1280

6. Waeber G, Calandra T, Roduit R et al (1997) Insulin secretion is regulated by the glucose-dependent production of islet beta cell macrophage migration inhibitory factor. Proc Natl Acad Sci U S A 94:4782–4787

7. Serre-Beinier V, Toso C, Morel P et al (2010) Macrophage migration inhibitory factor deficiency leads to age-dependent impairment of glucose homeostasis in mice. J Endocrinol 206:297–306

8. Nikolic I, Vujicic M, Saksida T et al (2013) The role of endogenous glucocorticoids in glucose metabolism and immune status of MIF-deficient mice. Eur J Pharmacol 714 (1–3):498–506

9. Vujicic M, Senerovic L, Nikolic I et al (2014) The critical role of macrophage migration inhibitory factor in insulin activity. Cytokine 69(1):39–46

10. Ohkawara T, Nishihira J, Takeda H et al (2002) Amelioration of dextran sulfate sodium-induced colitis by anti-macrophage migration inhibitory factor antibody in mice. Gastroenterology 123:256–270

11. Ohkawara T, Mitsuyama K, Takeda H et al (2008) Lack of macrophage migration inhibitory factor suppresses innate immune response in murine dextran sulfate sodium-induced colitis. Scand J Gastroenterol 43:1497–1504

12. Man AL, Lodi F, Bertelli E et al (2008) Macrophage migration inhibitory factor plays a role in the regulation of microfold (M) cell-mediated transport in the gut. J Immunol 181:5673–5680

13. Tlaskalová-Hogenová H, Stepánková R, Hudcovic T et al (2004) Commensal bacteria (normal microflora), mucosal immunity and chronic inflammatory and autoimmune diseases. Immunol Lett 93:97–108

14. Vujicic M, Saksida T, Despotovic S et al (2018) The role of macrophage migration inhibitory factor in the function of intestinal barrier. Sci Rep 8(1):6337

15. Yan Y, Kolachala V, Dalmasso G et al (2009) Temporal and spatial analysis of clinical and molecular parameters in dextran sodium sulfate induced colitis. PLoS One 4(6):e6073

16. Martinovic T, Ciric D, Pantic I et al (2018) Unusual shape and structure of lymphocyte nuclei is linked to hyperglycemia in type 2 diabetes patients. Tissue Cell 52:92–100

17. Graham L, Orenstein JM (2007) Processing tissue and cells for transmission electron microscopy in diagnostic pathology and research. Nat Protoc 2(10):2439–2450

18. Schindelin J, Arganda-Carreras I, Frise E et al (2012) Fiji: an open-source platform for biological-image analysis. Nat Methods 9 (7):676–682

Chapter 18

Investigating MIF in Mouse Models of Severe Corticosteroid-Resistant Neutrophilic Asthma

Venkata Sita Rama Raju Allam and Maria B. Sukkar

Abstract

Experimental mouse models of asthma are widely used to investigate the underlying mechanisms of this complex and heterogeneous disease. Using mouse models of ovalbumin-induced asthma, previous investigators have established a crucial role for MIF in the development of type 2-mediated eosinophilic asthma. Surprisingly, however, the role of MIF in other phenotypes of asthma has received little attention. MIF is an important mediator of neutrophilic inflammation, and also acts to antagonize the actions of corticosteroids. Thus, MIF may play a role in the development of severe forms of asthma in which airway neutrophilia and corticosteroid insensitivity are major features. In this chapter, we provide an experimental protocol that may be used to investigate the role of MIF in a mouse model of severe corticosteroid-resistant neutrophilic asthma.

Key words Severe asthma, Corticosteroid resistance, Neutrophil, Mouse model, House dust mite

1 Introduction

Macrophage migration inhibitory factor (MIF) was first identified as a mediator of asthma about 20 years ago. Studies published in the late 1990s reported increased levels of MIF protein in the bronchoalveolar lavage fluid (BALF), induced sputum, and serum of asthmatic subjects [1, 2]. In the subsequent decade, several investigators using *Mif* gene-deficient mice, neutralizing MIF antibodies, or the small-molecule MIF antagonist ISO-1 clearly demonstrated a role for MIF in mouse and rat models of ovalbumin (OVA)-induced asthma, which mostly recapitulate eosinophilic sub-phenotypes of asthma. In these studies, MIF inhibition protected against type 2-mediated eosinophilic inflammation, structural remodeling of the airway wall, and airway hyperreactivity, all of which are classic features of the allergic asthmatic response [3–8].

Over the past decade or so, it has become increasingly apparent, however, that asthma is a heterogeneous disease. Indeed, asthma

James Harris and Eric F. Morand (eds.), *Macrophage Migration Inhibitory Factor: Methods and Protocols*,
Methods in Molecular Biology, vol. 2080, https://doi.org/10.1007/978-1-4939-9936-1_18,
© Springer Science+Business Media, LLC, part of Springer Nature 2020

incorporates a spectrum of respiratory conditions which appear to be mediated by distinct and overlapping molecular mechanisms. This spectrum includes a group of patients who do not achieve or maintain control of their symptoms, despite adequate and, in many cases, high-dose treatment with anti-inflammatory corticosteroids. Defined as having severe asthma, this group constitutes about 5–10% of the asthmatic population, but accounts for nearly 50% of the health care costs of asthma, thus contributing a significant health and social burden [9].

MIF is a critical regulator of the immune response, and has the capacity to antagonize the anti-inflammatory effects of corticosteroids [10]. It is also a potent mediator of neutrophilic inflammation [11, 12]. MIF facilitates neutrophil recruitment to sites of inflammation, including the lung [13, 14], and prolongs neutrophil survival [15]. The increased recruitment of neutrophils to the airways in asthmatic subjects is one of the most important determinants of disease severity, and the lack of therapeutic response to corticosteroids [16, 17]. Thus, given that MIF regulates neutrophil trafficking and survival, and inhibits corticosteroid activity, anti-MIF therapies may offer a unique advantage in severe asthma, where neutrophilic inflammation and the lack of corticosteroid responsiveness present major therapeutic issues. In this chapter, we provide a detailed experimental protocol that may be used to investigate the role of MIF in a mouse model of severe corticosteroid-resistant neutrophilic asthma.

2   Materials

2.1   Induction of Severe Corticosteroid-Resistant Neutrophilic Asthma in Mice

1. C57BL6 female mice 8–12 weeks of age (*see* **Note 1**).

2. Isoflurane anesthetic system.

3. Mortar and pestle.

4. 1 mL Sterile syringe.

5. 25-gauge Needle.

6. 20 μL Pipette and tips.

7. 0.1% House dust mite (*Dermatophagoides pteronyssinus*) extract.

8. 1× Dulbecco's phosphate-buffered saline (PBS) (Life Technologies).

9. Complete Freund's adjuvant (CFA) (Sigma Aldrich).

10. 4,5-Dihydro-3-(4-hydroxyphenyl)-5-isoxazoleacetic acid methyl ester (ISO-1) (Tocris Bioscience).

11. Isoflurane (Veterinary Companies of Australia Pty Ltd).

12. 9α-Fluoro-16α-methyl-11β,17α,21-trihydroxy-1,4-pregna-diene-3,20-dione (Dexamethasone) (Sigma Aldrich).

13. 80% Ethanol.

2.2 Measurement of Airway Reactivity

1. flexiVent System (SCIREQ) (*see* **Note 2**),

2. 18-gauge Blunted cannula.

3. Spring scissors.

4. Fine scissors, sharply angled.

5. Two sets of small curved forceps.

6. Non-sterile silk thread.

7. Dissection board.

8. Absorbent paper towels.

9. Anesthetic agent (e.g., 1 mL of ketamine (100 mg/mL):0.1 mL xylazine (100 mg/mL):8.9 mL sterile water cocktail solution).

10. 10% Acetyl-β-methyl choline chloride (methacholine) (Sigma Aldrich).

11. 1x Dulbecco's phosphate-buffered saline (PBS) (Life Technologies).

12. 80% Ethanol.

2.3 Euthanasia and Collection of Biological Specimens

1. 23-gauge Needle.

2. Spring scissors.

3. Fine scissors.

4. Forceps.

5. Dissection board.

6. Absorbent paper towels.

7. 1.5 mL Eppendorf tubes.

8. Portable liquid nitrogen container.

9. Ice bucket.

10. Liquid nitrogen.

11. Pentobarbitone sodium (Lethabarb) (Virbac Animal Health).

12. 1× Hanks' balanced salt solution (HBSS).

13. RNA*later*® (Life Technologies).

14. 10% Neutral buffered formalin.

2.3.1 Bronchoalveolar Lavage Fluid (BALF) Processing

1. Refrigerated benchtop centrifuge (with rotor to accommodate 2 mL tubes).

2. 1.5 mL Eppendorf tubes.

3. 1 mL Pipette and tips.

4. 1× HBSS.

5. Red blood cell lysis buffer.

2.3.2 Blood Processing

1. Refrigerated benchtop centrifuge (with rotor to accommodate 2 mL tubes).

2. 1.5 mL Eppendorf tubes.

3. 1 mL Pipette and tips.

2.3.3 Cell Slide Preparation

1. Microscope (10×, 20×, 100× objectives).

2. Hemocytometer.

3. Coverslips.

4. Cytospin.

5. Shandon™ Single Cytofunnel™ with White Filter Cards.

6. Glass slides.

7. 1.5 mL Eppendorf tubes.

8. 1 mL Pipette and tips.

9. 20 µL pipette and tips.

10. Trypan blue.

11. Differential Quik Stain Kit (Modified Giemsa) (Polysciences, Inc.)

3   Methods

3.1 Induction of Severe Corticosteroid-Resistant Neutrophilic Asthma in Mice

To induce severe corticosteroid-resistant neutrophilic asthma, C57BL/6 mice are sensitized with the clinically relevant allergen house dust mite (HDM) in the presence of complete Freund's adjuvant (CFA), and then 14 days later challenged with a single dose of house dust mite. This protocol elicits profound airway neutrophilia, mucus hypersecretion, and airway hyperresponsiveness [18].

3.1.1 Preparation of House Dust Mite Extract and CFA Emulsion for Mouse Sensitization

1. Dissolve lyophilized house dust mite extract in PBS to achieve a concentration of 1 µg/1 µL (0.1%), and then transfer to a mortar which is to be kept on ice.

2. Slowly add an equal amount of CFA to the mortar.

3. Pulverize in a clockwise direction using the pestle (*see* **Note 3**).

4. After continuous pulverizing, an emulsion of house dust mite extract in CFA will form; this is used to sensitize mice as described below.

3.1.2 Allergen Sensitization via Subcutaneous Injection

1. Fill a 1 mL syringe with the HDM/CFA emulsion, and then fit the syringe with a 25-gauge needle.

2. Inject 100 μL of the HDM/CFA emulsion subcutaneously. Note, control mice are injected with PBS only.

3. Monitor mice for the development of lesions every second day (*see* **Note 4**).

3.1.3 Allergen Challenge via the Intranasal Route

1. 14 days after sensitization, place mice in the induction chamber of an isoflurane anesthetic system.

2. Adjust the oxygen flow rate to 1–2 L/min and the isoflurane vaporizer to 3% (*see* **Note 5**).

3. Mice will become immobilized and lightly sedated after approximately 2–3 min; this indicates that a suitable plane of anesthesia has been reached.

4. Once anesthesia is achieved, manually restrain the mouse in the supine position, with the head elevated.

5. Administer 100 μg HDM in a volume of 20 μL PBS into the nares using a 20 μL pipette. Note, control mice receive PBS only.

6. Once allergen administration is completed, mice can be immediately returned to their respective cages.

3.1.4 Administration of ISO-1 and Dexamethasone

1. To confirm corticosteroid-resistant airway neutrophilia, administer the corticosteroid dexamethasone (1 mg/kg) via oral gavage using a 20-gauge stainless steel curved feeding needle 1 h prior to allergen challenge (*see* **Note 6**).

2. To inhibit airway neutrophilia, mucus hypersecretion, and AHR, administer the MIF inhibitor ISO-1 (35 mg/kg) via intraperitoneal injection using a 25-gauge needle 1 h prior and 6 h post intranasal HDM administration. Control mice are injected with vehicle only.

3.2 Measurement of Airway Reactivity Using the flexiVent System

Prior to measuring airway reactivity, prepare all reagents and calibrate the flexiVent system. Following this, anesthetize the mice, cannulate the trachea, and measure airway reactivity as described below. Detailed instructions on how to measure airway reactivity using the flexiVent system have been previously published [19, 20].

3.2.1 Preparation of Reagents

1. Prepare the anesthetic agent by mixing 1 mL ketamine (100 mg/mL), 0.1 mL of xylazine (100 mg/mL), and 8.9 mL sterile water for injection in a 15 mL conical tube (*see* **Note 7**).

2. Prepare a stock solution of 10% methacholine by dissolving 100 mg of methacholine in 1 mL PBS. Perform serial dilutions

in PBS to generate concentrations of 50, 25, 10, 5, 2.5, 1.25 mg/mL or less, if needed (*see* **Note 8**).

3.2.2 Calibration of the flexiVent System

1. Weigh the mouse.
2. Enter the weight in the flexiVent operating software.
3. Proceed with the calibration of the system using the appropriate endotracheal cannula, according to the manufacturer's instructions (*see* **Note 9**).

3.2.3 Preparation of Mice

1. Anesthetize mice by injecting ketamine:xylazine cocktail 100 µL/10 g via the intraperitoneal route.
2. Return mice to their respective cages until they are sedated; this takes approximately 10–15 min.
3. Evaluate the animal's toe pinch reflex. A lack of response to the toe pinch indicates that the mouse has reached the surgical plane of anesthesia.
4. Place adsorbent paper towels on the dissection board, and then place the anesthetized mouse in a supine position on the dissection board, under a heat lamp to maintain body temperature.
5. Wipe down the fur on the neck area with 80% ethanol.
6. Make a vertical incision, starting between the forelimbs near the lower jaw, such that the submaxillary glands become visible.
7. To expose the trachea, separate the lobes of the submaxillary glands, and then, using fine scissors, dissect the sheath of muscular tissue that surrounds the trachea.
8. Place a 7–10 cm piece of thread underneath the trachea by holding the thread with forceps and dragging it below the trachea through to the other side. Once the thread is properly positioned, using spring scissors, make a small incision between cartilaginous rings on the trachea.
9. Insert an 18-gauge blunted cannula into the incision. Slide the cannula past the incision to make sure there is an adequate length of cannula inside the trachea. Tie the thread around the trachea in order to seal the cannula very tightly (*see* **Note 10**).

3.2.4 Measurement of Airway Reactivity

1. Start the flexiVent ventilator by selecting a predefined ventilation profile.
2. Connect the cannula to the ventilator system via the Y-tubing (*see* **Note 11**).
3. Run a deep inflation perturbation by double-clicking on the perturbation name to verify there is no leakage between the cannula insertion and Y-tubing (*see* **Note 12**).

4. Initiate a predefined script for measurement of airway reactivity by double-clicking on the script title in the software to run the automated measurement sequences.

5. Perform a baseline measure of airway reactivity in the absence of methacholine or PBS to ensure there are no aberrations in the trachea cannulation.

6. Load the nebulizer with PBS or methacholine, then activate the nebulizer to generate an aerosol for inhalation.

7. Measure airway reactivity to PBS and increasing concentrations of methacholine.

3.3 Euthanasia, Collection, and Processing of Blood, Bronchoalveolar Lavage Fluid, and Lung Tissue

1. Once measurements of airway reactivity are complete, stop the ventilator and detach the mouse from the flexiVent system.

2. Euthanize the mouse with pentobarbital (100 mg/kg) via the intraperitoneal route.

3. Blood samples, if required, must be collected immediately after euthanasia.

4. To obtain blood samples via cardiac puncture, place the mouse in a supine position on the dissection board covered with absorbent paper towels. Use a pair of fine scissors to open the chest cavity, being careful not to damage the lungs. Then carefully remove the collar bones, being careful not to damage the trachea. Following this, insert a 21- or 23-gauge needle into the heart and collect the blood. It is possible to collect approximately 600–700 μL of blood using this method.

5. Allow the blood to coagulate at room temperature for at least 30 min, then spin in a refrigerated centrifuge at $18,800 \times g$ for 5 min. Then carefully remove tubes from the centrifuge and collect the serum (which separates out as a distinct top layer) using a 1 mL pipette, and transfer to a new microcentrifuge tube. Store serum samples at $-80\ °C$ for later analysis.

6. If bronchoalveolar lavage fluid (BALF) is required, attach the hub of a syringe to the cannula that is attached to the trachea and slowly inject 1 mL of HBSS into the lungs in a uniform motion. Then immediately withdraw the HBSS from the lungs in a slow and continuous motion. The collected fluid can be stored in a 1.5 mL Eppendorf tube on ice for later processing, as described in **step 7**.

7. To pellet cells, spin the BALF in a refrigerated centrifuge at $900 \times g$ for 5 min. Carefully remove tubes from the centrifuge, being careful not to disturb the cell pellet, and transfer the supernatant to a clean 1.5 mL Eppendorf tube. Store BALF samples at $-80\ °C$ for later analysis. Then add 500 μL of red blood cell lysis buffer to the remaining cell pellet, and incubate on ice for 10 min. Spin cells in a refrigerated centrifuge at

        2500 × g for 5 min. Carefully remove tubes from the centrifuge, being careful not to disturb the pellet. Discard the supernatant, and resuspend cells in 1 mL HBSS.

8. To determine total cell counts, dilute 20 µL of the cell suspension in a 1:1 ratio with trypan blue. Transfer 20 µL of the cell suspension to a hemocytometer and count cells under a microscope with 10× magnification.

9. To perform differential cell counts, vortex the remaining 1 mL cell suspension at low speed. Load 150 µL of cell suspension onto labeled double-frosted microscope slides that have been secured into Shandon™ Single Cytofunnel™ with white filter cards, and cytospin at 600 rpm for 5 min. Allow slides to air-dry overnight, and then stain using Differential Quik Stain Kit (Modified Giemsa) according to the manufacturer's instructions. Allow stained slides to air-dry overnight. Perform differential cell counts using a microscope equipped with a 100× oil-immersed objective lens (*see* **Notes 13** and **14**).

10. The left lung lobe may be collected for histological analysis, while the right lung may be collected for measurement of gene and protein expression, if required, as described in **step 11** below.

11. To begin this process, ligate the right lung lobe using a suture thread. Then, with a second suture thread, support the trachea by placing the thread directly under it, below the end of the cannula, and loop the thread in a half-tightened knot. Once this is done, inflate the left lung lobe with 600 µL of 10% neutral buffered formalin. Be careful not to overinflate the lung, as this will distort tissue histology. Once the left lobe is fully inflated, tighten the half-tied suture thread, and remove the syringe and cannula from the trachea. Then, dissect out the right lung lobe, place it in a 1.5 mL Eppendorf tube, and immediately snap-freeze in liquid nitrogen. If the right lung lobe is required for gene expression studies, add an appropriate volume of RNA*later*® to the 1.5 mL Eppendorf tube. Then, to collect the left lung lobe, grasp the excess suture thread with the forceps and gently lift the trachea. Carefully excise the left lung lobe, with the heart still attached, and place the inflated left lobe in a 50 mL conical tube containing 50 mL of 10% neutral buffered formalin

4   Notes

1. All procedures are performed in living animals; thus, investigators must obtain appropriate approvals for working with animals at their institution, prior to conducting experiments.

2. Other alternatives are available for measurement of airway reactivity.

3. Ensure the emulsion is pulverized in one direction, preferably clockwise. Pulverization should be performed quickly in an uninterrupted manner until the emulsion forms. If not pulverized correctly, the emulsion will break and lose its physical stability.

4. Subcutaneous injection of the allergen/CFA emulsion may result in local inflammation and cause granulomatous lesions, skin ulceration, and abscesses at the site of injection. It is therefore important to monitor mice regularly, as serious lesions may require attention.

5. Exposure to isoflurane may occur during filling of the isoflurane anesthetic system, or during anesthesia, if the isoflurane escapes to the atmosphere. Exposure to isoflurane may cause severe respiratory distress and medullary paralysis; thus, proper use of the isoflurane anesthetic system and the waste scavenging system is essential.

6. To perform the oral gavage, restrain the mouse tightly by holding the scruff of its neck so that its head and body are in a straight vertical line. This allows easy passage of the feeding needle into the esophagus.

7. Measurement of airway reactivity in each mouse will take approximately 40–50 min. To maintain anesthesia for the duration of this period, administer additional anesthetic (one-third the original ketamine/xylazine dose) after 30 min have elapsed.

8. Methacholine is highly hygroscopic; fresh solutions must be prepared on the day of use.

9. The flexiVent system need only be calibrated at the start of each day. However, it is necessary to perform tube calibration prior to measuring airway reactivity in each individual mouse on any given day.

10. If necessary, adjust the position of the cannula to avoid aberrant breathing patterns during ventilation.

11. Position the animal as straight as possible on the flexiVent stand to ensure direct flow of air from the ventilator to the lungs. The mouse may be held in position by restraining the limbs with masking tape. When ready to proceed with mechanical ventilation, start the ventilator before attaching the cannula to avoid any interruption of airflow to the lungs.

12. To confirm there is no leakage around the cannula, deep inflation perturbation should be performed prior to commencing measurements of airway reactivity. This also allows standardization of lung volumes across different mice.

13. To ensure optimal results, slides for differential cell counts should be prepared the same day cells are retrieved from the mouse. It is also important that slides are completely dry prior to staining with Differential Quik Stain Kit (Modified Giemsa).

14. Eosinophils, neutrophils, and macrophages can be distinguished on the basis of several features; eosinophils have a lobed blue nucleus and cytoplasmic granules that stain reddish/orange, neutrophils have a lobed dark blue nucleus and unstained cytoplasmic granules, macrophages have a large round violet-stained nucleus and a sky blue cytoplasm.

References

1. Rossi AG et al (1998) Human circulating eosinophils secrete macrophage migration inhibitory factor (MIF). Potential role in asthma. J Clin Invest 101(12):2869–2874

2. Yamaguchi E et al (2000) Macrophage migration inhibitory factor (MIF) in bronchial asthma. Clin Exp Allergy 30(9):1244–1249

3. Mizue Y et al (2005) Role for macrophage migration inhibitory factor in asthma. Proc Natl Acad Sci U S A 102(40):14410–14,415

4. Wang B et al (2006) Cutting edge: deficiency of macrophage migration inhibitory factor impairs murine airway allergic responses. J Immunol 177(9):5779–5784

5. Kobayashi M et al (2006) Role of macrophage migration inhibitory factor in ovalbumin-induced airway inflammation in rats. Eur Respir J 27(4):726–734

6. Chen P-F et al (2010) ISO-1, a macrophage migration inhibitory factor antagonist, inhibits airway remodeling in a murine model of chronic asthma. Mol Med 16(9–10):400–408

7. Magalhães ES et al (2007) Macrophage migration inhibitory factor is essential for allergic asthma but not for Th2 differentiation. Eur J Immunol 37(4):1097–1106

8. Amano T, Nishihira J, Miki I (2007) Blockade of macrophage migration inhibitory factor (MIF) prevents the antigen-induced response in a murine model of allergic airway inflammation. Inflamm Res 56(1):24–31

9. Bittar HET, Yousem SA, Wenzel SE (2015) Pathobiology of severe asthma. Annu Rev Pathol Mech Dis 10(1):511–545

10. Calandra T et al (1995) MIF as a glucocorticoid-induced modulator of cytokine production. Nature 377:68

11. Makita H et al (1998) Effect of anti-macrophage migration inhibitory factor antibody on lipopolysaccharide-induced pulmonary neutrophil accumulation. Am J Respir Crit Care Med 158(2):573–579

12. Galvão I et al (2016) Macrophage migration inhibitory factor drives neutrophil accumulation by facilitating IL-1β production in a murine model of acute gout. J Leukoc Biol 99(6):1035–1043

13. Santos LL et al (2011) Macrophage migration inhibitory factor regulates neutrophil chemotactic responses in inflammatory arthritis in mice. Arthritis Rheum 63(4):960–970

14. Takahashi K et al (2009) Macrophage CD74 contributes to MIF-induced pulmonary inflammation. Respir Res 10(1):33

15. Baumann R et al (2003) Macrophage migration inhibitory factor delays apoptosis in neutrophils by inhibiting the mitochondria-dependent death pathway. FASEB J 17(15):2221–2230

16. Moore WC et al (2014) Sputum neutrophil counts are associated with more severe asthma phenotypes using cluster analysis. J Allergy Clin Immunol 133(6):1557–63.e5

17. Ray A, Kolls JK (2017) Neutrophilic inflammation in asthma and association with disease severity. Trends Immunol 38(12):942–954

18. Liu C et al (2017) The flavonoid cyanidin blocks binding of the cytokine interleukin-17A to the IL-17RA subunit to alleviate inflammation in vivo. Sci Signal 10(467): eaaf8823

19. McGovern TK et al (2013) Evaluation of respiratory system mechanics in mice using the forced oscillation technique. J Vis Exp 75: e50172–e50172

20. Hartney JM, Robichaud A (2013) Assessment of airway hyperresponsiveness in mouse models of allergic lung disease using detailed measurements of respiratory mechanics. Methods Mol Biol 1032:205–217

Chapter 19

Investigating MIF in Mouse Model of Gout

Izabela Galvão, Allysson Cramer, and Flavio Almeida Amaral

Abstract

Mice are widely used to assess the pathogenesis of diseases. An experimental model of gout consists of the injection of uric acid crystals into joints of mice, which reproduce inflammation and functional changes of the human disease. Uric acid crystals activate synoviocytes culminating in the release of IL-1β and neutrophil recruitment, key inflammatory elements in gouty arthritis. Since MIF plays an important role in orchestrating gout inflammation, we detail valuable procedures to investigate uric acid crystal-induced joint inflammation in mice and give options for further understanding the functions of MIF in gouty arthritis in vivo and in vitro.

Key words Gout, Mouse model of gout, Arthritis, MIF

1 Introduction

Gout is the most common inflammatory arthropathy, characterized by deposition of uric acid crystals in the joint leading to painful neutrophilic inflammation accompanied by edema and erythema. The canonical pathway associated with uric acid crystal-induced inflammation is the assembly of NLRP3 inflammasomes in synovial cells, activating caspase-1 and promoting IL-1β maturation and release [1]. Uncontrolled gouty inflammation and hyperuricemia contribute to recurrent gout flares, which can affect multiple joints and progressively cause chronification due to urate deposition in soft tissue (tophi), progressive joint destruction, chronic pain, and other complications, such as renal failure [2].

Much of our knowledge about the mechanisms related to gouty inflammation come from proof-of-concept models in animals. The injection of monosodium urate (MSU) crystals into the joint of mice induces inflammatory and functional changes locally. This mouse model displays a fair representation of the human disease. The injection of MSU crystals induces a rapid accumulation of neutrophils in the joint cavity, which is elicited by pro-inflammatory chemokines (CXCL1 and CXCL2) and

James Harris and Eric F. Morand (eds.), *Macrophage Migration Inhibitory Factor: Methods and Protocols*,
Methods in Molecular Biology, vol. 2080, https://doi.org/10.1007/978-1-4939-9936-1_19,
© Springer Science+Business Media, LLC, part of Springer Nature 2020

cytokines (IL-1β and TNF-α). Neutrophils contribute to inflammatory pain, IL-1β production, and tissue damage, all related to gouty disability [3]. It should be noted that this is an acute model, which means that the inflammation resolves around 24 h after injection of MSU crystals, associated with apoptosis of the accumulated neutrophils, reduction of pro-inflammatory mediators, and restoration of tissue damage [4]. In particular, rodents express uricase, which accelerates MSU crystal degradation [5].

Macrophage migration inhibitory factor (MIF) plays a significant role in the development of chronic inflammatory conditions [6]. It is a pro-inflammatory cytokine produced by macrophages in response to a variety of inflammatory stimuli [7]. MIF regulates leukocyte trafficking by binding to chemokine receptors and increasing chemokine expression [8]. Migrated neutrophils also release MIF, maintaining a positive feedback on neutrophil recruitment in different systems [9]. We have previously shown that MIF plays a major role in MSU crystal-induced joint inflammation in mice via its ability to control local IL-1β production [10]. MIF can be easily identified in tissues by different methods after the induction of gout in mice. Thus, we provide strategies to investigate the functions of MIF in vivo and in vitro in MSU-induced arthritis in mice by evaluating joint tissues and primary cell cultures.

2 Materials

2.1 MSU Crystal Preparation

1. Uric acid solution (add 1.68 g of uric acid diluted in 500 mL of NaOH 0.01 M solution).
2. Cell culture filter, 0.45 μm.
3. Ethanol PA.
4. Phosphate-buffered saline (PBS), sterile.

2.2 Fibroblast-Like Synoviocytes Culture

1. 70 % Ethanol.
2. Laminar flow cabinet.
3. Complete medium: DMEM with 15% FBS, 1% penicillin/streptomycin (5 mg/mL stock).
4. Fetal bovine serum.
5. Scissors and tweezers washed and decontaminated with 70% ethanol.
6. Petri dish.
7. 0.5% Collagenase IV solution: 10 mg of collagenase IV dissolved in 2 mL of complete medium and filtered with a 0.22 μm syringe filter. Prepare the solution just before use.
8. Syringe, 3 mL.
9. 0.22 μm Syringe filter.

10. Orbital shaker.

11. Trypsin/EDTA (2.5 g/L/250 mg/L, stock solution).

12. Incubator.

2.3 Gout Model

1. 8–12-week-old male mice.

2. Syringe preparation is described in ref. 11.

2.4 MIF Expression in Synovium Tissue

1. RNA lysis buffer (e.g., TRizol).

2. Commercial reverse transcriptase kit.

3. Commercial master mix.

4. 500 nM of primer for *mif*, 5′-CAGAACCGCAACTACAGT AAGC-3′ (forward) and 5'-GGTGGATAAACACAGAACACT ACG-3′ (reverse) and for gapdh, 5'-ACG GCC GCA TCT T CT TGT GCA-3′ (forward) and 5'-CGG CCA AAT CCG TTC ACA CCG A-3′ (reverse).

2.5 Knee Wash

1. PBS/BSA 3%.

2. Formaldehyde 3.7%.

3. FC block commercially available (e.g., Anti-CD16/32).

4. PBS/BSA 1%.

5. Antibody anti-MIF.

6. Fluorescently conjugated secondary antibodies.

2.6 MIF Measurement in Synoviocytes

1. MIF levels in the supernatant detected using commercial kit.

2. MIF expression using cell lysate buffer to protein extraction or total RNA extraction.

3 Methods

3.1 Preparation of MSU Crystals

1. Add 1.68 g of uric acid diluted in 500 mL of NaOH 0.01 M solution. The solution has white color and should have pH around 6–6.5.

2. Add NaOH to reach the solution pH between 7.1 and 7.2 and heat at 70 °C until dissolved. This step usually takes a while due to the buffered solution.

3. Once completely dissolved and translucent, filter the solution with a cell culture filter (0.45 μm) and keep it sterile (*see* **Note 1**).

4. Transfer the solution to a sterile plastic tube or glass flask. Leave the solution at room temperature stirring slowly until you see crystal formation. Glass flasks normally precipitate crystals faster, around 3–5 days. Crystals should form on the wall of the flask/tube.

5. Discard supernatant and wash the crystals in ethanol. At this step, the crystals are big and have nonuniform size.

6. Sonicate the crystals in ethanol for 30 min to decrease their size.

7. Centrifuge ($400 \times g$ for 10 min), discard ethanol, and dry the crystals. Resuspend the crystals in PBS and make aliquots before use—adapted from ref. 1.

8. It is fundamental to perform the steps in sterile conditions. The analysis of endotoxin contamination must be done before the use of MSU crystals.

3.2   Gout Model

The acute gout model involves the injection of MSU crystals (100 μg/10 μL) into the tibiofemoral cavity of 8–12-week-old male mice. Contralateral joint could be used as a negative control (injection of PBS—MSU crystal diluent). The preparation of the syringe and the injection into the joint are described in [11].

Injection into the joint:

1. Once anesthetized (according to local ethical committee), remove the fur over the knee. This is easily done by pulling the fur with your fingers.

2. Slightly flex the knee in order to facilitate the injection.

3. Use a needle or a nail to find the exact location of the articular cavity.

4. Draw 10 μL of the 100 μg MSU crystal into the syringe. This volume is sufficient to fill all the tibiofemoral cavity of the mouse.

5. Inject the volume when the syringe level is perpendicular to the knee.

3.3   Measurement of MIF in Periarticular Tissue

1. After euthanasia, the skin from the knee is removed to expose the joint.

2. Periarticular tissue is removed from the joint using scalpel and forceps (Fig. 1). The tissue must be stored at −20 °C until processed for the analysis.

3. Homogenize the tissue in PBS containing anti-proteases in a proportion of 1 mL of solution to 100 mg of tissue.

4. Centrifuge the homogenate at 10,000 rpm for 10 min at 4 °C and collect the supernatant for determination of MIF concentration by ELISA (*see* **Note 2**).

3.4   Evaluation of MIF in Synovial Tissue

1. After euthanasia, remove the skin from the knee and cut the patellar tendon to expose the joint cavity.

2. Collect the synovial tissue using scalpel and fine point forceps (Fig. 2). The synovia is small, so it is necessary to pool at least

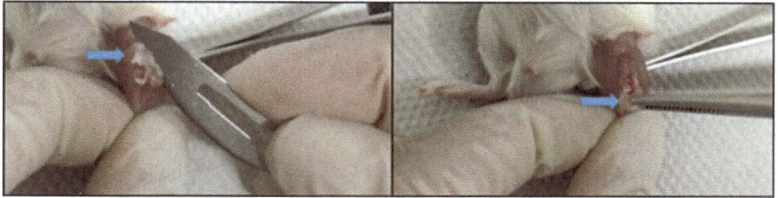

Fig. 1 Periarticular tissue is the tissue that is around the joint cavity

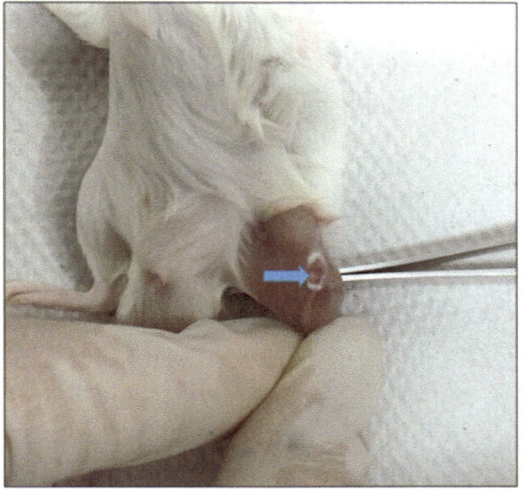

Fig. 2 Synovial tissue is collected inside the articular cavity, between femur and tibia

two synovia in the same tube (*see* **Note 3**). Store at −80 °C until analysis.

3. Homogenize the synovial tissue in RNA lysis buffer for extraction of total RNA following by cDNA synthesis using commercial reverse transcriptase Kit.

4. Perform real-time quantitative PCR using commercial master mix of choice.

5. 10 ng of cDNA from the samples is normally used in each reaction.

6. The thermal cycling conditions used are: an initial DNA denaturation step at 95 °C for 10 min, followed by 40 cycles of denaturation at 95 °C for 15 s, primer annealing, and extension at 60 °C for 1 min. Finally, melting curve analysis is performed by slowly cooling the PCRs from 95 to 60 °C (0.05 °C per cycle) with simultaneous measurement of the signal intensity (e.g., SYBR green). Melting-point determination analysis allowed the confirmation of the specificity of the amplification products.

7. Relative expression of the *Mif* genes is determined by $\Delta\Delta Ct$ method, using the constitutive gene, *Gapdh*, used as an internal control [10].

3.5 Identification of MIF in Macrophage Recovered from Knee Lavage

1. After euthanasia, remove the skin from the knee and cut the patellar tendon to expose the joint cavity.

2. Cut the synovial membrane, wash the tibiofemoral cavity using 5 μL of 3% BSA in PBS three times, and add to a tube containing 90 μL of 3% BSA in PBS. Repeat this step.

3. Cytospin 5×10^5 cells from the lavage onto slides (*see* **Note 4**).

4. Fix the cells/slides in 3.7% formaldehyde for 30 min.

5. After fixation, incubate with FC block for 30 min.

6. Stain the cells with macrophage surface marker (e.g., F4/80, CD11b, and CD68 [12]) and incubate overnight.

7. Wash the cells with 1% BSA in PBS and permeabilize cells using an appropriate commercial cell permeabilization kit (*see* **Note 5**).

8. Incubate with antibody against MIF overnight.

9. Wash cells with 1% BSA in PBS and incubate with appropriate fluorescently conjugated secondary antibodies for 30 min.

10. Images can be obtained with a confocal microscope. Fluorescence intensity can be measured off-line using appropriate software.

3.6 Fibroblast-Like Synoviocyte Purification

1. Use at least three mice per group. Inject MSU crystals (100 μg; 10 μL) into the tibiofemoral joint (i.a.), as in Subheading 3.2 above.

2. Control mice receive an i.a. injection of PBS only.

3. At selected time points after MSU challenge, euthanize the mice prior to dissection of inflamed joints.

4. Rinse microdissecting instruments with 70% ethanol and then immerse for the decontamination after cutting the joints of each mouse.

5. Immerse killed mouse in the 70% ethanol for 2 min.

6. In a laminar flow hood, remove the skin using scissors and tweezers and cut the patellar tendon over the knee.

7. Use a scalpel to remove the synovia (Fig. 3) and transfer to a petri dish with complete medium (*see* **Note 6**).

8. Prepare collagenase IV solution and filtrate in 0.22 μm syringe filter.

9. Add the synovium tissue collected from 3 mice in 2 mL of complete medium + collagenase IV in a sterile plastic tube. Add synovia from triplicate mice for each group to the same tube.

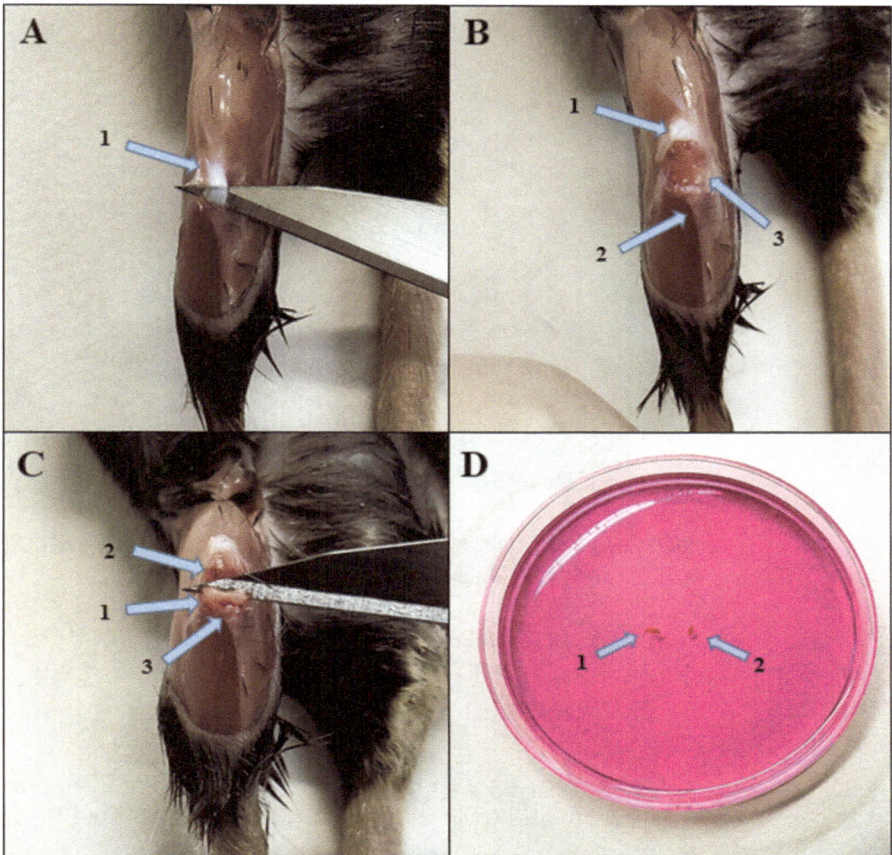

Fig. 3 Separation and isolation of synovial tissue from C57BL/6 mice. (**a**) Cut the patellar tendon (1) with the aid of a scalpel. (**b**) Behind the patellar tendon (1) and above the tibial bone (2) is located the synovium (3). (**c**) The synovium (1) is carefully removed with a scalpel, cutting vertically between the femur (2) and the tibia (3). (**d**) The collected synovium tissue is transferred to a petri dish containing DMEM 15% SFB

10. Incubate the synovia with collagenase solution in an orbital shaker at 37 °C for 1 h at 120 rpm. At the end of the incubation, use a vortex to vigorously release the cells.

11. Transfer the suspension of the digested synovium to a fresh 15 mL tube and add 10 mL of complete medium and centrifuge the tubes at $300 \times g$ for 10 min at room temperature.

12. Discard the supernatant and resuspend the pellet in 10 mL of complete medium.

13. Transfer the suspension (3 mouse or 6 synovia) to a 25 cm^2 flask and incubate at 37 °C, 5% CO_2.

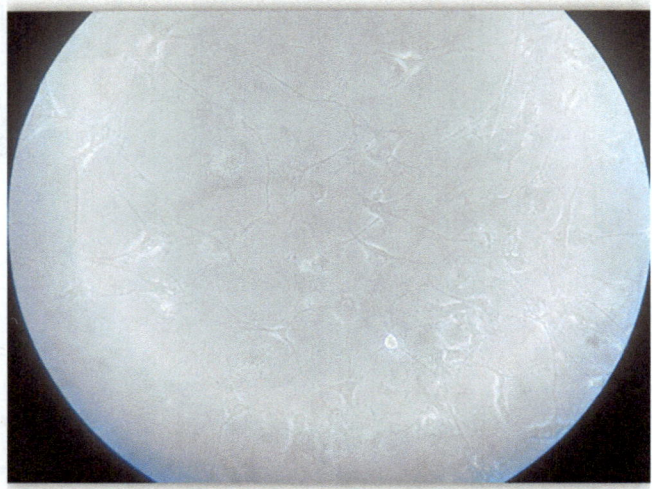

Fig. 4 Culture of fibroblast-like synoviocytes. Cells after 25 days of culture in the flask (25 cm^2) with DMEM supplemented with 15% of FBS.

3.7 Culture and Utilization of Fibroblast-Like Synoviocytes

1. Isolated cells take about 5–7 days to adhere.

2. After observing the adherence of the cells in the 25 cm^2 flask (5–7 days), change complete medium every 3 days (*see* **Note** 7).

3. After reaching a confluence of approximately 90%, remove the medium and wash the flasks with 5 mL of sterile PBS at 37 °C.

4. Remove PBS with serological pipette and add 2 mL of trypsin/EDTA (from stock solution) per flask. Incubate for 5 min in the culture incubator (37 °C, 5% CO$_2$).

5. Add 10 mL of complete medium and centrifuge at 300 × *g* for 10 min at room temperature, then transfer to a 75 cm^2 flask.

6. The first confluence typically takes approximately 20–25 days (Fig. 4).

7. After the first pass repeat the passage procedure in the ratio of 1:2 (the confluence is reached faster; from 10 to 15 days approximately).

8. After the 3rd passage, the synovial-like fibroblasts are ready to use. Frequently, these cells are used from the third to the ninth passage.

3.8 Identification of MIF in Fibroblast-Like Synovial Cells

1. MIF in the supernatant can be quantified by ELISA after different time points following MSU crystal stimulation.

2. Cell lysate can be obtained for total RNA extraction to perform qPCR or for protein extraction (cell lysis buffer) to perform Western Blotting assay.

4   Notes

1. Keep sterile all the steps during the procedures for MSU crystals preparation.

2. For homogenization of periarticular tissue, make sure that all the tissue was completely homogenized. After the centrifugation, the supernatant should be clear. The supernatant can be stored at -20 °C for later measurement of molecules of interest.

3. The synovial tissue collection should be made carefully using a sharp scalpel. The tissue is too tiny and sensitive. After cutting the tissue, collect it with the forceps.

4. During cytospin preparation, make sure that all slides were clean enough; otherwise, any artifact can disturb the image in the confocal microscopy.

5. Cell permeabilization can be performed; instead, use a commercial kit, using two types of reagents: organic solvents such as methanol and acetone or detergent such as Triton X-100 (0.1%) and saponin (0.01%).

6. During removal of synovia, it is important to avoid cutting bones with the scalpel to prevent contamination by bone marrow cells.

7. During the first 2 weeks of fibroblast-like synoviocytes, keep small pieces of synovium adhered to the bottle.

8. To investigate the role of MIF it is possible to use some inhibitors. Treatment with neutralizing antibody systemically or with small molecule to blockage the MIF receptor locally (intra-articular injection) are good strategies [10].

Acknowledgement

We thank Ilma Marçal for technical assistance. We thank the funding agencies Conselho Nacional de Desenvolvimento Cientifico e Tecnológico (CNPq, Brazil), Fundação de Amparo à Pesquisa do Estado de Minas Gerais (FAPEMIG, Brazil), and Coordenação de Aperfeiçoamento de Pessoal de Nível Superior (CAPES) for financial support.

References

1. Martinon F et al (2006) Gout-associated uric acid crystals activate the NALP3 inflammasome. Nature 440(7081):237–241

2. So AK, Martinon F (2017) Inflammation in gout: mechanisms and therapeutic targets. Nat Rev Rheumatol 13(11):639–647

3. Amaral FA et al (2012) NLRP3 inflammasome-mediated neutrophil

recruitment and hypernociception depend on leukotriene B(4) in a murine model of gout. Arthritis Rheum 64(2):474–484

4. Galvao I et al (2017) Annexin A1 promotes timely resolution of inflammation in murine gout. Eur J Immunol 47(3):585–596

5. Johnson RJ, Lanaspa MA, Gaucher EA (2011) Uric acid: a danger signal from the RNA world that may have a role in the epidemic of obesity, metabolic syndrome, and cardiorenal disease: evolutionary considerations. Semin Nephrol 31(5):394–399

6. Morand EF, Leech M, Bernhagen J (2006) MIF: a new cytokine link between rheumatoid arthritis and atherosclerosis. Nat Rev Drug Discov 5(5):399–410

7. Santos LL et al (2008) Reduced arthritis in MIF deficient mice is associated with reduced T cell activation: down-regulation of ERK MAP kinase phosphorylation. Clin Exp Immunol 152(2):372–380

8. Gregory JL et al (2006) Macrophage migration inhibitory factor induces macrophage recruitment via CC chemokine ligand 2. J Immunol 177(11):8072–8079

9. Daryadel A et al (2006) Apoptotic neutrophils release macrophage migration inhibitory factor upon stimulation with tumor necrosis factor-alpha. J Biol Chem 281(37):27653–27661

10. Galvao I et al (2016) Macrophage migration inhibitory factor drives neutrophil accumulation by facilitating IL-1beta production in a murine model of acute gout. J Leukoc Biol 99 (6):1035–1043

11. Amaral FA, Boff D, Teixeira MM (2016) In vivo models to study chemokine biology. Methods Enzymol 570:261–280

12. Zhang X, Goncalves R, Mosser DM (2008) The isolation and characterization of murine macrophages. Curr Protoc Immunol., Chapter 14:Unit 14 1

Chapter 20

5′ and 3′ RACE Method to Obtain Full-Length 5′ and 3′ Ends of *Ciona robusta* Macrophage Migration Inhibitory Factors *Mif1* and *Mif2* cDNA

Aiti Vizzini

Abstract

The 5′ and 3′ RACE is a method to obtain full-length 5′ and 3′ ends of cDNA using known cDNA sequences from expressed sequence tags (ESTs), subtracted cDNA, differential display, or library screening. Here is described the identification of full-length 5′ and 3′ ends of *Ciona robusta Mif1* and *Mif2* cDNA by using 5′ and 3′ RACE method.

Key words 5′ and 3′ RACE, cDNA, Genome, *Ciona robusta*

1 Introduction

Rapid amplification of complementary DNA (cDNA) ends (RACE) is a powerful technique for obtaining the ends of cDNAs when only partial sequences are available [1, 2]. It is especially useful in the studies of temporal and spatial regulation of transcription initiation and differential splicing of mRNA. The 5′ and 3′ RACE method ensures the amplification of only full-length transcripts via elimination of truncated messages from the amplification process. With the GeneRacer® Kit (Invitrogen) full-length 5′ and 3′ ends can be cloned, to construct complete cDNA sequences [3]. The system described provides a set of prequalified reagents intended for synthesis of first-strand cDNA, purification of first-strand products, and preparation of target cDNA for subsequent amplification by PCR. The method described in this chapter to identify full-length 5′ and 3′ ends of *Ciona robusta Mif1* and *Mif2* cDNA [3, 4] is quite efficient. RACE PCR products can be quickly and easily cloned using either the Zero Blunt® TOPO® PCR Cloning Kit for Sequencing (blunt-end PCR products) or the TOPO TA Cloning® for Sequencing Kit.

James Harris and Eric F. Morand (eds.), *Macrophage Migration Inhibitory Factor: Methods and Protocols*,
Methods in Molecular Biology, vol. 2080, https://doi.org/10.1007/978-1-4939-9936-1_20,
© Springer Science+Business Media, LLC, part of Springer Nature 2020

2 Materials

1. Coraliquid marine invertebrate food (Sera, Heinsberg, Germany).
2. RNAlater Tissue Collection solution.
3. RNAqueous Midi Kit purification system.
4. Total RNA (4 μg) in DEPC water (approximately 0.5–1 μg/μL).
5. 1.5 mL Sterile microcentrifuge tubes.
6. Heat block set at 50 °C.
7. 95% Ethanol, 70% ethanol.
8. Microcentrifuge at room temperature and 4 °C.
9. Sterile water: Sterile, diethylpyrocarbonate (DEPC)-treated ("DEPC water").
10. GeneRacer® Kit (Invitrogen), all the components in the list below, from (a) to (g), are included in the kit:

    (a) RNaseOut™, calf intestinal phosphatase (CIP), tobacco acid pyrophosphatase (TAP), T4 RNA ligase, cloned avian myeloblastosis virus reverse transcriptase (cloned AMV RT)

    (b) 10× CIP buffer: 0.5 M Tris–HCl, pH 8.5 (20 °C)1 mM EDTA

    (c) 10× TAP buffer: 0.5 M Sodium acetate, pH 6.0, 10 mM EDTA, 1% β-mercaptoethanol, 0.1% Triton® X-100

    (d) 10× T4 RNA ligase buffer: 330 mM Tris-acetate, pH 7.8 (25 °C), 660 mM potassium acetate,100 mM magnesium acetate, 5 mM DTT

    (e) 10 mM ATP mix

    (f) Random primers

    (g) 5× Reverse transcription buffer: 250 mM Tris acetate (pH 8.4), 375 mM potassium acetate, 40 mM magnesium acetate, stabilizer 20 μg/mL BSA

    (h) Phenol/Chloroform solution: Phenol:chloroform:isoamyl alcohol (25:24:1), 0.1% 8-hydroxyquinoline; mussel glycogen: 10 mg/mL in DEPC water; 3 M sodium acetate in DEPC water, pH 5.2; 14. 0.1 M DTT in DEPC water; RNase H: 2 U/μL in: 20 mM Tris–HCl, pH 7.5; 100 mM KCl, 10 mM MgCl$_2$, 0.1 mM EDTA, 0.1 mM DTT, 50 μg/mL BSA, 50% glycerol; dNTP Mix (10 mM each); Random Primers (N6) 100 ng/μL in DEPC water (54 μM); GeneRacer™ Oligo dT Primer 820 ng/μL in

DEPC water (50 μM); 100 mM dNTPs in 200 mM Tris–HCl, pH 7.5.

11. DNA gel: UltraPure Agarose 2% in TBE buffer.

12. 100 bp DNA Ladder.

3 Methods

3.1 Preparation of Total RNA

In our study [4], Ascidians were collected from Sciacca Harbor (Sicily, Italy), maintained in tanks with aerated seawater at 15 °C, and fed with Coraliquid marine invertebrate food (Sera Heinsberg, Germany). Before excision of fragments of pharynx, the tunic surface was cleaned and sterilized with ethyl alcohol. Fragments of pharynx tissue (200 mg) were immediately soaked in RNAlater Tissue Collection solution and stored at −80 °C. The pharynx occupies an extensive part of the adult body beneath the tunic surface. It consists of two epithelial monolayers perforated by dorsoventrally aligned rows of elongated elliptical, ciliated stigmata enclosed in a mesh of vessels (also called transversal and longitudinal bars), where the hemolymph, containing abundant mature and immature hemocytes, flows. Hemopoietic nodules are associated with the bar epithelia. Total RNA extraction was performed using a RNAqueous Midi Kit purification system (*see* **Notes 1** and **2**).

3.2 Preparation of cDNA RACE

This technique is based on RNA ligase-mediated (RLM-RACE) and oligo-capping rapid amplification of cDNA ends (RACE) methods and results in the selective ligation of an RNA oligonucleotide to the 5′ ends of decapped mRNA using T4 RNA ligase [4–7]. This technique consists of four phases: dephosphorylating RNA with calf intestinal phosphatase (CIP) to remove the 5′ phosphates, removing the mRNA cap structure, ligating the RNA oligo to decapped mRNA, reverse transcribing mRNA with cloned AMV RT (Fig. 1).

3.2.1 Dephosphorylating RNA

1. Assemble 10 μL of dephosphorylation reaction components on ice in a 1.5 mL sterile microcentrifuge tube: RNA (4 μg), the volume depending on the concentration; 1 μL 10× CIP buffer;1 μL RNaseOut™ (40 U/μL); 1 μL CIP (10 U/μL); DEPC water adds to reach 10 μL.

2. Mix gently by shaking and centrifuge to collect fluid to bottom. Incubate at 50 °C for 1 h.

3. After incubation, centrifuge briefly and place on ice.

4. To precipitate RNA, add 90 μL DEPC water and 100 μL phenol/chloroform solution and vortex vigorously.

5. Centrifuge at 11,000 rpm or 15,500 × g in a microcentrifuge for 5 min at room temperature.

Dephosporylation with CIP

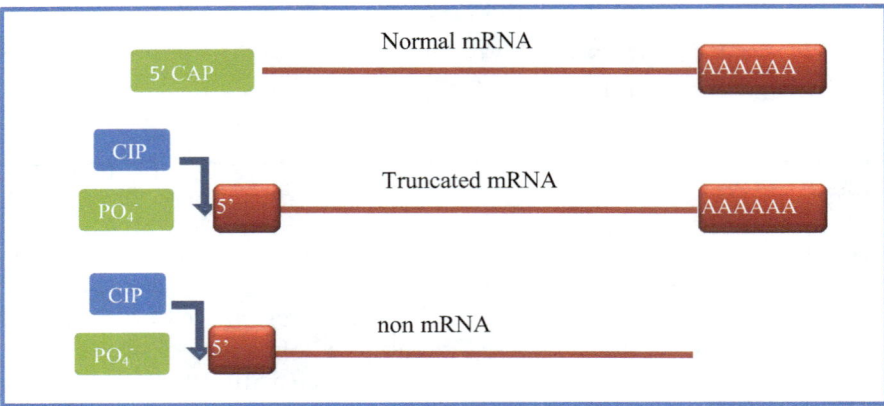

Decapped mRNA with TAP

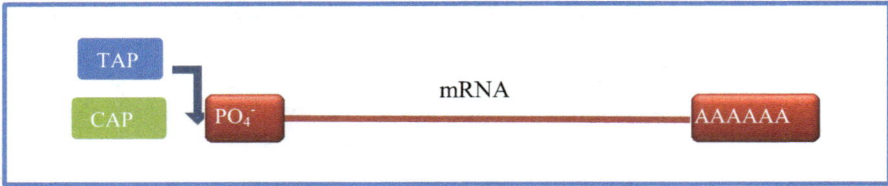

Ligation of GeneRacer RNA oligo

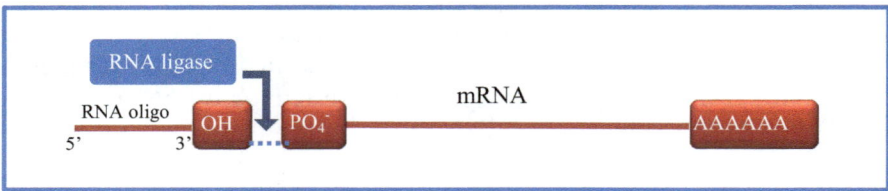

Reverse transcription

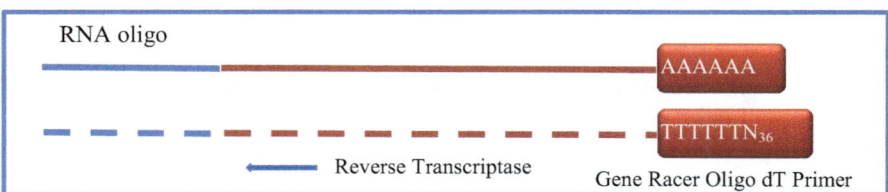

Fig. 1 cDNA RACE method scheme

6. Transfer aqueous (top) phase to a new microcentrifuge tube (~100 μL).

7. Add 2 μL 10 mg/mL mussel glycogen, 10 μL 3 M sodium acetate, pH 5.2, and mix well. Add 220 μL 95% ethanol and vortex briefly. Freeze on dry ice for 10 min.

8. To pellet RNA, centrifuge at 11,000 rpm or 15,500 × g in a microcentrifuge for 20 min at 4 °C.

9. Remove 95% ethanol by pipet, avoiding resuspension of the pellet.

10. Add 500 μL 70% ethanol, invert several times, and vortex briefly.

11. Centrifuge at 11,000 rpm or 15,500 × g in a microcentrifuge for 2 min at 4 °C, remove the ethanol by pipet avoiding to resuspend the pellet.

12. Centrifuge again for 1 min at 11,000 rpm or 15,500 × g to collect remaining ethanol. Carefully remove the remaining ethanol by pipet and air-dry the pellet for 1–2 min at room temperature.

13. Finally resuspend the pellet in 7 μL DEPC water.

3.2.2 Removing the mRNA Cap Structure

After dephosphorylating and precipitating the RNA, you are ready to remove the 5′ cap structure from full-length mRNA.

1. Assemble 10 μL of decapping reaction components on ice in a 1.5 mL sterile microcentrifuge tube; 7 μL dephosphorylated RNA; 1 μL of 10× TAP buffer; 1 μL RNaseOut™ (40 U/μL); 1 μL TAP (0.5 U/μL).

2. Mix gently by shaking. Centrifuge at 11,000 rpm or 15,500 × g briefly to collect fluid. Incubate at 37 °C for 1 h. After incubation, centrifuge briefly and place on ice.

3. To precipitate RNA, add 90 μL DEPC water and 100 μL phenol/chloroform solution and vortex vigorously. Centrifuge at 11,000 rpm or 15,500 × g in a microcentrifuge for 5 min at room temperature.

4. Transfer aqueous (top) phase to a new microcentrifuge tube (~100 μL).

5. Add 2 μL 10 mg/mL mussel glycogen, 10 μL 3 M sodium acetate, pH 5.2, and mix well. Add 220 μL 95% ethanol and vortex briefly. Freeze on dry ice for 10 min.

6. To pellet RNA, centrifuge at 11,000 rpm or 15,500 × g in a microcentrifuge for 20 min at 4 °C.

7. Remove 95% ethanol by pipet avoiding to resuspend the pellet.

8. Add 500 μL 70% ethanol, invert several times, and vortex briefly.

9. Centrifuge at 11,000 rpm or 15,500 × *g* in a microcentrifuge for 2 min at 4 °C, remove the ethanol by pipet avoiding resuspension of the pellet.

10. Centrifuge again for 1 min at 11,000 rpm or 15,500 × *g* to collect remaining ethanol. Carefully remove the remaining ethanol by pipet and air-dry the pellet for 1–2 min at room temperature.

11. Resuspend the pellet in 7 μL DEPC water.

3.2.3 Ligating the RNA
Oligo to Decapped mRNA

1. Add 7 μL of dephosphorylated decapped RNA to the tube containing the pre-aliquoted, lyophilized GeneRacer™ RNA Oligo (0.25 μg). Pipet up and down several times to mix and resuspend RNA Oligo. Centrifuge briefly to collect the fluid in the bottom of the tube.

2. Incubate at 65 °C for 5 min to relax the RNA secondary structure. After the incubation, the total volume of this solution may decreased by 1 μL due to evaporation.

3. Place on ice to chill (~2 min) and centrifuge briefly.

4. Assemble 10 μL of Ligase reaction: 6 μL of dephosphorylated decapped RNA and GeneRacer™ RNA Oligo; 1 μL 10× Ligase Buffer; 1 μL 10 mM ATP; 1 μL RNaseOut™ (40 U/μL); 1 μL T4 RNA ligase (5 U/μL).

5. Mix gently by shaking. Centrifuge briefly to collect fluid. Incubate at 37 °C for 1 h. After incubation, centrifuge briefly and place on ice.

6. To precipitate RNA, add 90 μL DEPC water and 100 μL phenol/chloroform solution and vortex vigorously. Centrifuge at 11,000 rpm or 15,500 × *g* in a microcentrifuge for 5 min at room temperature.

7. Transfer aqueous (top) phase to a new microcentrifuge tube (~100 μL).

8. Add 2 μL 10 mg/mL mussel glycogen, 10 μL 3 M sodium acetate, pH 5.2, and mix well. Add 220 μL 95% ethanol and vortex briefly. Freeze on dry ice for 10 min.

9. To pellet RNA, centrifuge at 11,000 rpm or 15,500 × *g* in a microcentrifuge for 20 min at 4 °C.

10. Remove 95% ethanol by pipet avoiding resuspension of the pellet.

11. Add 500 μL 70% ethanol, invert several times, and vortex briefly.

12. Centrifuge at 11,000 rpm or 15,500 × *g* in a microcentrifuge for 2 min at 4 °C, remove the ethanol by pipet avoiding to resuspend the pellet.

13. Centrifuge again for 1 min at 11,000 rpm or $15,500 \times g$ to collect remaining ethanol. Carefully remove the remaining ethanol by pipet and air-dry the pellet for 1–2 min at room temperature.

14. Resuspend the pellet in 10 μL DEPC water.

3.2.4 Reverse Transcribing mRNA with Cloned AMV RT

1. Add 1 μL of the random primer and 1 μL of dNTP Mix (25 mM each) to the ligated RNA (10 μL).

2. Incubate at 65 °C for 5 min to remove any RNA secondary structure.

3. Chill on ice for 2 min and centrifuge at 11,000 rpm or $15,500 \times g$ briefly.

4. Assemble 20 μL of reverse transcription reaction: 12 μL ligated RNA and primers obtained from point 3; 4 μL 5× RT buffer; 1 μL cloned AMV RT (15 U/μL); 2 μL sterile water; 1 μL RNaseOut™ (40 U/μL).

5. Mix well and incubate the reaction mix at 25 °C for 10 min and then incubate the reaction mix at 45 °C for 1 h.

6. Incubate at 85 °C for 15 min to inactivate cloned AMV RT.

7. Centrifuge at 11,000 rpm or $15,500 \times g$ briefly and use immediately for amplification or store at −20 °C.

3.3 Preparation of Partial Sequences of Mif1 and Mif2 cDNA and Design of Gene-Specific (GSP) Primers for 5′ 3′ PCR RACE

We previously designed GSP primers based on sequence, amplified, and sequence identified in the Ciona genome: *Mif1* (ENSCING00000021369) and *Mif2* (ENSCING00000021423) [4]. To amplify and sequence the fragments of *Mif1* and *Mif2* cDNA the primer pairs were used for amplifying cDNA obtained by reverse transcription of mRNA of pharynx fragments excised from the body wall. PCR was performed using the primers listed in Table 1. Amplification comprised 2 min of initial denaturation at 95 °C, followed by 30 cycles consisting of 95 °C for 30 s, 1 min at 60 °C, 72 °C for 1 min, and a final extension at 72 °C for 7 min. A single band, respectively, of 231 bp and 248 bp in size was detected. Finally, PCR products were cloned and sequenced.

To perform 5′ and 3′ RACE you need GSP and GSP NESTED primers with the following characteristics:

1. 50–70% GC content to obtain a high annealing temperature (>72 °C).

2. 23–28 nucleotides in length to increase specificity of binding.

3. Low GC content at 3′ ends to minimize extension by DNA polymerase at nontarget sites (no more than two G or C residues in the last five bases).

4. No self-complementary sequences within the primer or no sequence.

Table 1
Primers used for cloning

Gene	Primer sequence (5'–3')	Application
*Cr*Mif1 *Forward*	5'-ACCAACGTTTCAAGCGACAA-3'	PCR
*Cr*Mif1 *Reverse*	5'-AAGAGGAGGCATCAAAAGCG-3'	PCR
*Cr*Mif2 *Forward*	5'-AGACAAGTTGCCAAAGAGCA-3'	PCR
*Cr*Mif2 *Reverse*	5'-GTCTTGTGCAACTCCCAGTA-3'	PCR
*Cr*Mif1	5'-CGCTTCGGATTTACAAGAGATAGTGGT-3'	RACE5'
	5'-CGCAGCATCTCAACAAACCGAAAGCA-3'	NESTED5'
	5'-AGAGGAGGCATCAAAAGCGCTGC-3'	RACE3'
	5'-ACAATAGAAGTGACTGTACAGCAAGCA-3'	NESTED3'
*Cr*Mif2	5'-TGGTCTCTTCAACAATTCCAAACAAGCC-3'	RACE5'
	5'-TCTGTGTGACTGTGGTTCCTGATCTGT-3'	NESTED5'
	5'-TGTCTTGTGCAACTCCCAGTAACTTGT-3'	RACE3'
	5'-ACAAGGCTCCTCAGTCCCATCAAATGAC-3'	NESTED3'
GeneRacer 5' Primer	5'-CGACTGGAGCACGAGGACACTGA-3'	RACE5'
GeneRacer 5' Nested Primer	5'-GGACACTGACATGGACTGAAGGAGTA-3'	NESTED5'
GeneRacer 3' Primer	5'-GCTGTCAACGATACGCTACGTAACG-3'	RACE3'
GeneRacer 3' Nested Primer	5'-CGCTACGTAACGGCATGACAGTG-3'	NESTED3'

5. Complementary to the primers supplied in the kit, especially at the 3' end.

6. Annealing temperature greater than 72 °C. Using primers with a high annealing temperature will improve the specificity of your PCR and will allow you to use touchdown PCR (RACE PCR).

3.4 Amplification of the cDNA RACE

3.4.1 RACE PCR

Perform two PCR reactions in 0.5 tubes, respectively, with Gene RACER 5' and reverse GSP1, Gene RACER 3' primers and forward GSP1 listed in Table 1 (*see* also Fig. 2).

1. Assemble 50 μL of PCR reaction as follows:

   (a) 1 μL RT template, 3 μL GeneRacer™ 5' or 3' primer (10 μM).

   (b) 1 μL Reverse GSP (10 μM) or forward GSP (10 μM).

   (c) 10 μL 10× high fidelity PCR buffer.

   (d) 1 μL dNTP solution (10 mM each).

   (e) 0.5 μL Platinum® Taq DNA Polymerase High Fidelity (5 U/μL).

5'RACE

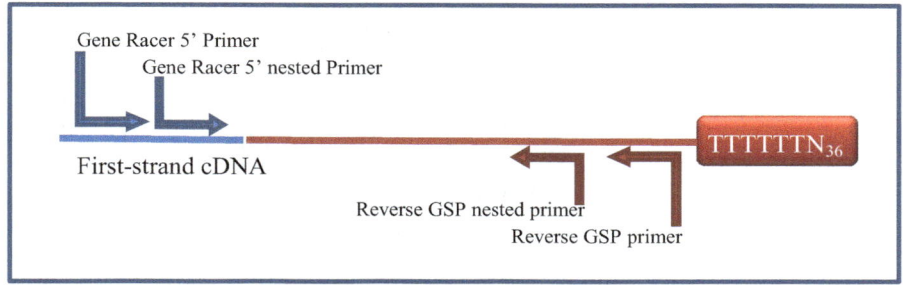

3'RACE

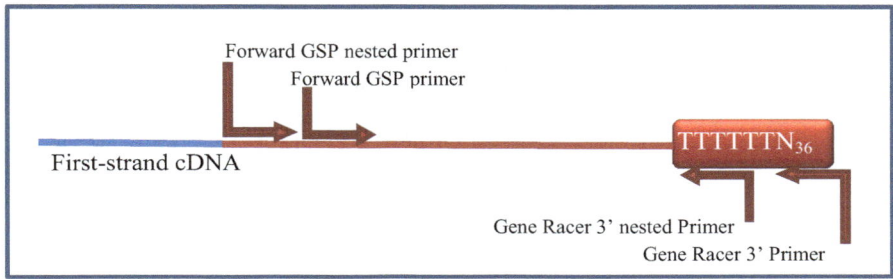

Fig. 2 PCR RACE scheme

  (f) 2 μL MgSO$_4$ (50 mM).

  (g) 36.5 μL Sterile water.

2. Amplification comprises 2 min of initial denaturation at 95 °C, followed by 5 cycles consisting of 94 °C for 30 s and 72 °C for 1 min, followed by 5 cycles consisting of 94 °C for 30 s and 70 °C for 1 min, followed by 25 cycles consisting of 94 °C for 30 s, 60 °C for 30 s, 72 °C for 1 min, and a final extension at 72 °C for 10 min.

3. After PCR, analyze 10 μL of the amplification reaction on a 1% agarose gel (Fig. 3a).

4. If you observe multiple bands or a smear, perform a Nested PCR (*see* **Notes 3** and **4**).

3.4.2 Nested PCR

1. Use 1 μL of the original amplification reaction (RACE PCR) as a template for nested PCR.

   Perform two PCR reactions in 0.5 tubes, respectively, with Gene RACER 5′ NESTED and reverse GSP1 NESTED, Gene RACER 3′ NESTED primers and forward GSP1 NESTED listed in Table 1 (Fig. 2). Assemble 50 μL of PCR reaction as follows:

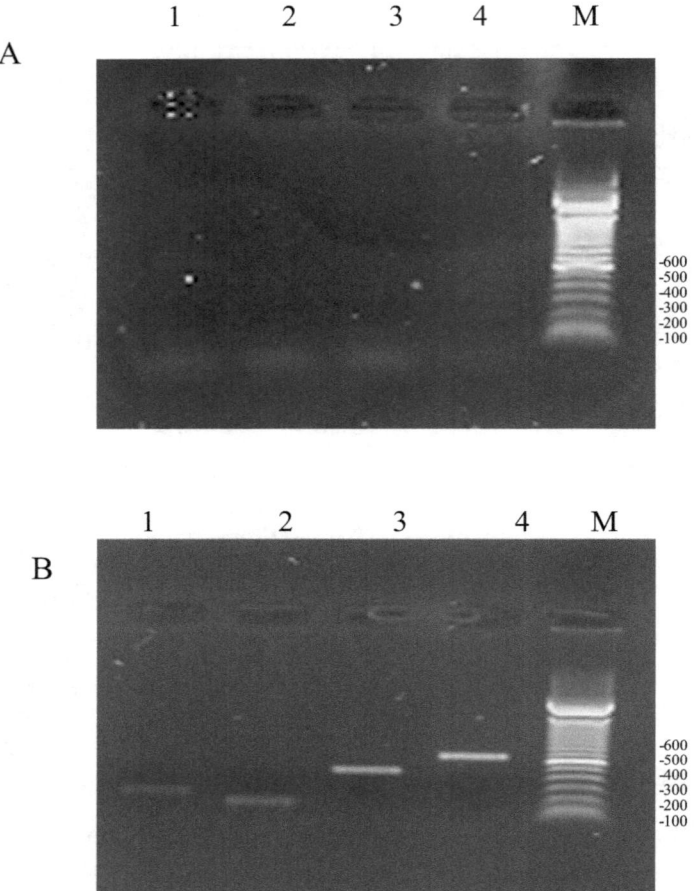

Fig. 3 PCR 5' and 3' RACE products of *C. robusta Mif1* and *Mif2*. (**a**) PCR 5' and 3' NESTED products of *C. robusta Mif1* and (**b**) *Mif2*

(a) 1 μL Template; μL GeneRacer™ 5' NESTED or GeneRacer™ 3' primer NESTED (10 μM).

(b) 1 μL Reverse GSP NESTED (10 μM) or forward GSP NESTED (10 μM); 5 μL 10× High Fidelity PCR Buffer.

(c) 1 μL dNTP solution (10 mM each).

(d) 0.5 μL Platinum® Taq DNA Polymerase High Fidelity (5 U/μL).

(e) 2 μL $MgSO_4$ (50 mM).

(f) 36.5 μL Sterile water.

2. Amplification comprises 2 min of initial denaturation at 95 °C, followed by 25 cycles consisting of 94 °C for 30 s, 60°Cfor 30 s, 72 °C for 2 min, and a final extension at 72 °C for 10 min.

3. After PCR, analyze 10 μL of the amplification reaction on a 1% agarose gel (Fig. 3b). Nested PCR products will be shorter by the number of bases between the original primers and the nested primers.

3.5 Cloning and Sequencing

1. Excise the band from the gel by Gel Extraction Kit.

2. Clone the overlapping RACE products into the pCR™IIvector (TA Cloning Kit, Invitrogen) and sequence. They contained the complete coding regions.

---

4 Notes

1. When working with RNA ensure that the RNA preparation is free of agents that inhibit reverse transcription. Use RNASe- and DNAse-free pipette tips with filter. Wear latex gloves while handling reagents and RNA samples to prevent RNase contamination from the surface of the skin.

2. The first step is to check the concentration of RNA obtained by using spectrophotometry or NanoDrop spectrophotometers. The concentration of an RNA solution can be determined by measuring its absorbance at 260 nm (A_{260}) using a spectrophotometer. With a traditional spectrophotometer, dilute an aliquot of the RNA 1:50–1:100 in TE (10 mM Tris–HCl pH 8, 1 mM EDTA), and read the absorbance (Be sure to zero the spectrophotometer with the TE used for sample dilution). NanoDrop spectrophotometers are more convenient—no dilutions or cuvettes are needed, just measure 1.5 μL of the RNA sample directly. To determine the RNA concentration in μg/mL, multiply the A_{260} by the dilution factor and the extinction coefficient (1 A_{260} = 40 μg RNA/mL).

$$A_{260} \times \text{dilution factor} \times 40 = \text{μg RNA/mL}$$

Be aware that any contaminating DNA in the RNA prep will lead to an overestimation of yield, since all nucleic acids absorb at 260 nm. RNA quality can be tested as follows:

(a) Spectrophotometry: an effective measure of RNA purity is the ratio of absorbance readings at 260 and 280 nm; it should fall in the range of 1.8–2.1.

(b) Electrophoreses: To check the RNA for integrity, analyze 500 ng of your RNA by agarose 1% gel electrophoresis. You may use a denaturing agarose gel: 5 mL of 10× MOPS running buffer, in 45 mL of deionized sterile water and use MOPS1X as running buffer [8]. Prepare the RNA sample with Formaldehyde Load Dye, denature samples at 68 °C for 5 min and in ice. Run the gel until the bromophenol blue (the faster-migrating dye) has migrated as far as 2/3 the length of the gel. For a total RNA of good quality, you should see the 28S and 18S rRNA bands clearly, the mRNA will appear as a smear from 0.5 to 12 kb. Discard the RNA preparation if ribosomal bands are not sharp.

(c) Microfluidic analysis: Microfluidic systems such as the Agilent® 2100 Bioanalyzer™ with Caliper's RNA Lab-Chip® Kits provide better quantitative data than conventional gel analysis for characterizing RNA. Using a Bioanalyzer™, the RIN (RNA Integrity Number) can be calculated to further evaluate RNA integrity. A metric developed by Agilent, the RIN analyzes information from both rRNA bands, as well as information contained outside the rRNA peaks (potential degradation products) to provide a fuller picture of RNA degradation states.

3. To have success in amplification you need to optimize the PCR by using a hot start and touchdown PCR to minimize the background. The hot start PCR method minimizes mispriming and extension and can be achieved using Platinum® Taq DNA Polymerase High Fidelity that provide an automatic hot start. Touchdown PCR increases specificity and reduces background amplification by starting at a high annealing temperature in order to amplify only gene-specific or GeneRacer™-tagged cDNA, allowing the desired product to accumulate, and decreasing the annealing temperature through the remaining PCR cycles permits efficient amplification of tagged, gene-specific template [9, 10].

4. Some potential problems can be smeared product or no specific amplification. To increase sensitivity and specificity you need to:

(a) Perform nested PCR.

(b) Reduce the amount of template or use two- to fourfold diluted RT template.

(c) Reduce the number of PCR cycles.

(d) Increase the annealing temperature to eliminate nonspecific primer binding.

References

1. Frohman MA, Dush MK, Martin GR (1988) Rapid production of full-length cDNAs from rare transcripts: amplification using a single gene-specific oligonucleotide primer. Proc Natl Acad Sci U S A 85:8998–9002

2. Zhang Y, Frohman MA (1997) Using rapid amplification of cDNA ends (RACE) to obtain full-length cDNAs. In: Cowell IG, Austin CA (eds) Methods in molecular biology, vol 69. Humana, Totowa, NJ, pp 61–87

3. Invitrogen (2000) Advanced RACE method amplifies only full-length cDNA ends. Expr Newsl 7(3):2–3

4. Vizzini A, Parisi MG, Di Falco F, Cardinale L, Cammarata M, Arizza V (2018) Identification of CPE and GAIT elements in 3′ UTR of macrophage migration inhibitory factor (MIF) involved in inflammatory response induced by LPS in *Ciona robusta*. Mol Immunol 99:66–74

5. Maruyama K, Sugano S (1994) Oligo-capping: a simple method to replace the cap structure of eukaryotic mRNAs with oligoribonucleotides. Gene 138:171–174

6. Schaefer BC (1995) Revolutions in rapid amplification of cDNA ends: new strategies for polymerase chain reaction cloning of full-length cDNA ends. Anal Biochem 227:255–273

7. Volloch V, Schweitzer B, Rits S (1994) Ligation-mediated amplification of RNA from murine erythroid cells reveals a novel class of beta-globin mRNA with an extended 5′-untranslated region. Nucleic Acids Res 22:2507–2511

8. Ausubel FM, Brent R, Kingston RE, Moore DD, Seidman JG, Smith JA, Struhl K (1994) Current Protocols in molecular biology. Greene Publishing Associates and Wiley-Interscience, New York

9. Don RH, Cox PT, Wainwright BJ, Baker K, Mattick JS (1991) "Touchdown" PCR to circumvent spurious priming during gene amplification. Nucleic Acids Res 19:4008

10. Roux KH (1995) Optimization and troubleshooting in PCR. PCR Methods Appl 4:5185–5194

Purification of Antibodies Against *Entamoeba histolytica* MIF and Their Use in Analyzing Human and Mouse Samples

Laura Farr, Koji Watanabe, and Shannon Moonah

Abstract

Macrophage migration inhibitory factor (MIF) is a proinflammatory and proproliferative cytokine expressed in humans. MIF homologs also exist in many pathogenic protozoans, including *Entamoeba*, *Plasmodium*, *Toxoplasma*, and *Leishmania*. Production of antibodies against parasite proteins allows for the generation of assays to measure and visualize parasite infection within hosts. In this chapter, we describe how to specifically purify antibodies against *Entamoeba histolytica* MIF (*Eh*MIF), and subsequently use anti-*Eh*MIF antibodies for ELISA on mouse and human samples and for immunohistochemistry on human tissue. These methods can be applied to any protein for high-quality antibody purification.

Key words Amebiasis, *Entamoeba*, Parasite

1 Introduction

Macrophage migration inhibitory factor (MIF) is a proinflammatory cytokine that mediates the innate immune response. Counterintuitively, MIF homologs also exist in pathogenic protozoans such as *Entamoeba histolytica*, *Plasmodium*, *Toxoplasma*, and *Leishmania* [1]. *Entamoeba histolytica* MIF (*Eh*MIF) induces damaging inflammation by exploiting the innate immune system to promote infiltration of host tissue. The adaptive immune system uses *Eh*MIF as an antigen, and anti-*Eh*MIF antibodies provide a protective effect against future infections [2, 3]. Detection of *Eh*MIF has been suggested as a diagnostic test for human amebiasis; *Eh*MIF is detected by stool ELISA, and infection is easily visualized through immunohistochemistry on biopsies [4].

To measure *Eh*MIF in mouse and human samples, a polyclonal antibody was generated by inoculating rabbits against recombinant *Eh*MIF. Western blot validation of antibodies purified by protein A resulted in multiple nonspecific bands. We suspected that we had nonspecific antibodies from the rabbit serum because protein A

James Harris and Eric F. Morand (eds.), *Macrophage Migration Inhibitory Factor: Methods and Protocols*, Methods in Molecular Biology, vol. 2080, https://doi.org/10.1007/978-1-4939-9936-1_21,

binds to the Fc portion of antibodies non-selectively. In pursuit of specific anti-*Eh*MIF antibodies, a literature search was performed on antigen-based purification of antibodies. Covalently binding recombinant proteins to a resin allows for the binding and subsequent elution of antibodies from serum without elution of the antigen. Preclearance of serum with the tag for the fusion protein used in generation and purification of the antibody also helps reduce the number of nonspecific antibodies produced *from contaminants from bacterial expression* systems unintentionally included during animal inoculation [5–7]. The purified antibody was subsequently used in ELISAs for human and mouse samples [2], and immunohistochemistry on both mouse [2, 8] and human tissue [4]. Importantly, this technique can be applied to any protein for generation and high-quality purification of custom antibodies.

2   Materials

2.1   GST and-GST-EhMIF Protein Expression

1. LB-Agar-Amp plates containing GST and GST-*Eh*MIF.
2. LB-Amp: LB Broth with 100 μg/mL Ampicillin (or appropriate antibiotic).
3. Shaking incubator, capable of temperatures 15 °C to 37 ° C and speeds of 220 rpm.
4. 10 mL Culture tubes.
5. 2 L Flasks.
6. 100 mM IPTG (isopropyl β-D-1-thiogalactopyranoside) in sterile DI water.

2.2   Antibody Purification

1. GST-tag fusion protein.
2. Glutathione Sepharose 4B Fast Flow.
3. Chromatography columns (10 mL).
4. 250 mM NaCl in PBS (recommended).
5. Polymyxin sulfate salt (0.1 mg/mL) in PBS (recommended).
6. 4–20% TGX Protein Gel (optional).
7. Coomassie Blue (optional).
8. 9.2 mM BS3 (bis(sulfosuccinimidyl)suberate, MW = 368.35) in 20 mM sodium phosphate, pH 7.9, 0.15 mM NaCl in dH_2O.
9. Tube rotator (end-over-end recommended).
10. 0.1 M Ethanolamine in PBS.
11. Reduced glutathione elution buffer, pH 7.8: 50 mM Tris–HCl pH 9, 250 mM NaCl, 50 mM L-glutathione reduced in dH_2O.
12. 0.1 M Glycine, pH 2.8 in dH_2O.
13. Amicon Ultra 0.5 Centrifugal Filter 3 kDa MWCO.

2.3 Alternative-NHS-Activated Agarose Antibody Purification (See Note 1)

1. MBP-tag fusion protein.
2. Phosphate-buffered saline, pH 7.4.
3. Amylose resin.
4. Amylose column buffer: 20 mM Tris, pH 7.4, 200 mM NaCl, 1 mM EDTA.
5. Amylose elution buffer: 10–50 mM Maltose in amylose column buffer.
6. Amicon Ultra 0.5 Centrifugal Filter 30 kDa MWCO.
7. NHS-activated agarose dry resin, 33 mg.
8. 1 M Tris–HCl in dH_2O, pH 7.4.
9. 1% (w/v) Sodium azide in PBS.
10. 0.1 M Glycine in dH_2O, pH 2.8.
11. 1 M Tris in dH_2O, pH 9.
12. Amicon Ultra 0.5 Centrifugal Filter 3 kDa MWCO.

2.4 ELISA

1. 96-Well high-protein-binding polystyrene plates.
2. Phosphate-buffered saline.
3. Anti-*Eh*MIF (5 µg/mL) made in Subheading 3.2.
4. 1% (w/v) Bovine serum albumin in PBS, pH 7.4.
5. PBS, pH 7.4 with 0.05% Tween-20.
6. Recombinant *Eh*MIF.
7. Biotinylated anti-*Eh*MIF: 0.25 µg/mL in 1% BSA in PBS, pH 7.4 (*see* **Note 6**).
8. Avidin-conjugated horseradish peroxidase.
9. 3, 3′, 5, 5′-Tetramethylbenzidine (ELISA detection reagent).
10. 1 M Phosphoric acid.

2.5 Serum Anti-EhMIF Detection

1. 96-Well high-protein-binding polystyrene plates.
2. Carbonate-bicarbonate buffer, pH 9.5.
3. 0.25% (w/v) Bovine serum albumin in PBS-0.05% Tween-20.
4. PBS, pH 7.4 with 0.05% Tween-20.
5. Recombinant *Eh*MIF.
6. HRP-conjugated anti-human total IgG: dilute 1:10,000 in 0.25% BSA in PBS-0.05% Tween-20.
7. 3, 3′, 5, 5′-Tetramethylbenzidine (TMB) ELISA detection reagent.
8. 1 M Phosphoric acid.

2.6 Immunohisto-
chemistry

1. Charged glass slides.

2. Target Retrieval Solution, high pH.

3. PT Link Instrument.

4. Autostainer.

5. Antibody diluent, ready-to-use diluent.

6. Peroxidase and alkaline phosphatase blocking reagent.

7. Envision Rabbit Link.

8. 3, 3'-Diaminobenzidine tetrachloride (DAB+).

9. Hematoxylin.

3   Methods

3.1 GST and GST-
EhMIF Protein
Expression (See
Note 1)

1. From LB-Agar-Amp plates, make 2.5 mL LB-Amp cultures from single colonies of GST and GST-*Eh*MIF. Incubate overnight in a shaking incubator at 37 °C, 220 rpm.

2. When turbid, add overnight cultures to 250 mL LB-Amp in 2. L flasks. To ensure enough yield, we normally grow 2 to 4 flasks of each protein at a time.

3. Shake the flasks at 220 rpm and 37 °C until the cells reach mid-log phase (OD600nm = 0.5–0.6).

4. Shift the temperature to 15 °C and induce the culture(s) with IPTG to a final concentration of 1 mM (2.5 mL of 100 mM IPTG stock solution per 250 mL of culture). Continue shaking at 220 rpm overnight for 18–24 h.

5. Recover the cells by centrifugation at $4000 \times g$ for 15 min at 4 °C. Store at −80 °C for future lysis or continue to Subheading 3.2.

3.2 Antibody
Purification
with a GST-EhMIF
Protein (Fig. 1a)

1. On ice, resuspend GST and GST-EhMIF pellets with 2.5 mL 1× phosphate-buffered saline, pH 7.4 (*see* **Note 2**). Lyse by sonication or other homogenization technique at 4 °C. Transfer lysate to 1.5 mL microtubes and centrifuge for 30 min at $>16,000 \times g$, 4 °C to remove the insoluble portion.

2. Equilibrate 0.5–1.5 mL of Glutathione Sepharose 4B Fast Flow (we recommend 500 μL bed volume per 250 mL culture) with PBS. In 15–50 mL tubes (based on cleared lysate volume), incubate cleared GST or GST-*Eh*MIF lysate with equilibrated glutathione sepharose overnight at 4 °C while rotating slowly or rocking. Optionally, if cleared lysates are concentrated, e.g., <5 mL from 250 mL culture, add an equal volume of PBS to assist binding.

3. Centrifuge at $1000 \times g$ and remove the lysate.

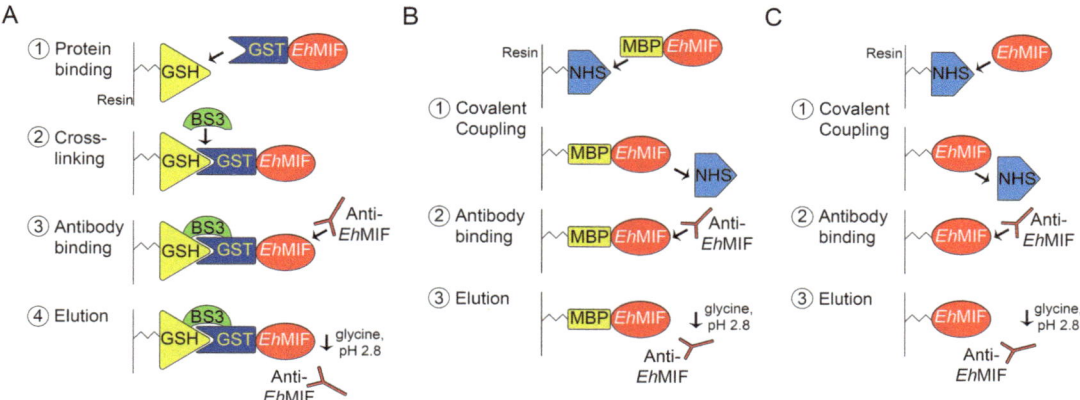

Fig. 1 Schematic illustration of anti-*Eh*MIF antibody purification using cross-linking with BS3 (**a**) and NHS-activated beads (**b** and **c**)

4. Resuspend the glutathione sepharose in 10 mL cold PBS and transfer to a 10 mL chromatography column. Allow column to wash via gravity flow. Optionally, wash tube with 5 mL of PBS, add to column, and repeat to ensure all sepharose is collected.

5. Subsequently wash with approximately 10 mL of the following solutions (in order): 1× PBS, 250 mM NaCl in 1× PBS, Polymyxin sulfate salt 0.1 mg/mL, and 1× PBS.

6. After the final wash, the sepharose column may be stored at 4 °C prior to testing and used for affinity purification.

7. Expression and binding of GST and GST-fusion protein may be confirmed by boiling samples from each column in Laemmli Sample Buffer, separating them on a 4–20% TGX gel (or SDS-PAGE gel at similar concentration, which is an appropriate substitute), and staining the gel with Coomassie blue.

8. Add 5 mL of 9.2 mM BS3 to column containing GST fusion protein bound to glutathione sepharose in amine-free buffer. Rotate (end-over-end to prevent clumping) for 45 min at room temperature. Drain by gravity flow.

9. Quench the reaction by adding 5 mL of 0.1 M ethanolamine [6, 7, 9] or 1 M Tris–HCl, pH 7.5 [6]; rotate end-over-end for 10 min to prevent clumping.

10. Drain the ethanolamine by gravity flow, and then wash with 5 mL of ethanolamine or 1 M Tris–HCl pH 7.5 used in **step 9**, allowing it to drain immediately by gravity flow.

11. Wash sepharose once with 1× PBS and rotate until the sepharose is completely resuspended before draining by gravity flow.

12. Elute any unbound GST proteins with 5 mL of fresh reduced glutathione elution buffer, pH 7.8.

13. Rotate end-over-end or mix for 10 min before draining.

14. Repeat glutathione elution once.

15. Optional pause point: Store columns at 4 °C until ready to purify serum.

16. Immediately prior to serum purification, elute any nonspecifically bound proteins from the cross-linked sepharose with glycine; add 5 mL 0.1 M glycine pH 2.8 to the column and incubate for 5 min while rotating or mixing before draining.

17. Repeat glycine elution once.

18. Wash column with 10–20 mL 1× PBS thoroughly via gravity flow to ensure all glycine is removed.

19. Add serum to column containing GST bound to glutathione sepharose. Rotate end-over-end for 1 h at room temperature or overnight at 4 °C.

20. Collect serum from GST column via gravity flow (*see* **Note 3**), adding PBS until sepharose returns to its original color.

21. Add serum to cross-linked GST-fusion protein column.

22. Rotate column end-over-end or mix for 2–3 h at room temperature or overnight at 4 °C. Collect serum via gravity flow, adding PBS until sepharose returns to its original color.

23. Elute antibodies with 1 mL 0.1 M glycine (pH 2.8) for every 2 mL serum originally added (*see* **Note 4**). Collect in 0.5–1 mL aliquots, and promptly neutralize aliquots with 1:10 1 M Tris–HCl, pH 9 (100 μL in each 1 mL aliquot immediately after collection).

24. Repeat glycine elution once.

25. Wash column with 10 mL PBS to ensure all glycine is removed and collect wash. Repeat as necessary. Column may be stored at 4 °C in PBS.

26. Pool all neutralized elutes of affinity-purified antibody (*see* **Note 5**) and dialyze into 1× PBS using 3 kDa centrifugal filter, 15 mL (Millipore Sigma). Briefly, add all elutes (and optionally first PBS wash) to centrifugal filter. Add PBS to 15 mL and centrifuge at $4000 \times g$ for 30–40 min or until 2 mL remain.

27. Add PBS to 15 mL and repeat centrifugation to 2 mL. May repeat once to reduce glycine concentration and concentrate antibody to 2 mL.

28. Check protein concentration, and validate antibody with immunoblot of recombinant protein.

3.3   Alternative NHS-Activated Agarose Antibody Purification (See Note 1 and Fig. 1)

1. Prior to this protocol, express both MBP and MBP-*Eh*MIF. For optimal yield, pellet cells, and resuspend with 1 mL amylose column buffer: 100 mL culture volume. Lyse with sonication, and centrifuge to obtain cleared lysate. In a column, add cleared lysate to 200 μL equilibrated amylose resin per 100 mL original culture volume, and add 1 mL amylose column buffer per 1 mL cleared lysate. Incubate overnight, rotating end-over-end at 4 °C.

2. Remove lysate via gravity flow. Wash with 10 bed volumes of amylose column buffer.

3. Add 1 bed volume of amylose column buffer to the MBP column and store it at 4 °C.

4. Elute MBP-*Eh*MIF from column by adding 2 bed volumes of amylose elution buffer. Incubate rotating end-over-end between 30 and 60 min.

5. Collect eluted MBP-*Eh*MIF and wash the column with 10 bed volumes of column buffer. Add 1 bed volume of column buffer to the MBP-*Eh*MIF column, and store at 4 °C.

6. Combine eluted MBP-*Eh*MIF and post-elution wash into a 30 kDa centrifugal filter and add amylose column buffer to 15 mL. Remove the maltose from the MBP-*Eh*MIF by centrifuging until 1 mL remains.

7. Buffer exchange MBP-*Eh*MIF into PBS (*see* **Note 6**) by adding 14 mL PBS to the centrifugal filter and centrifuging until 1 mL remains. Repeat addition of PBS and centrifugation.

8. Measure concentration of MBP-*Eh*MIF via preferred method. Optimally, produce at least 4 mg of MBP-*Eh*MIF in amine-free buffer.

9. Before coupling MBP-*Eh*MIF to NHS resin, set aside 1 mL of MBP-*Eh*MIF.

10. For 4 mg of protein, add two 33 mg columns of NHS-activated agarose dry resin to the MBP-*Eh*MIF. Rinse the columns with the MBP-*Eh*MIF set aside in **step 1** and add to the total mixture.

11. Rotate end-over-end for 2 h at room temperature.

12. Add the NHS resin columns (with the caps removed) to 1.5 mL microtubes. Transfer the resin and protein to the columns 500 μL at a time and centrifuge for 1 min at 1000 × *g* and collect flow-through to measure binding efficiency. Repeat transfer, rinsing tube with 400 μL PBS, until all resin is returned to the columns, collecting the flow-through.

13. Wash columns once with 400 μL PBS and collect flow-through.

14. Neutralize the coupling reaction by adding 400 µL of 1 M Tris, pH 7.4 to each column. Cap each end and rotate end-over-end for 20 min at room temperature.

15. Remove caps and add columns to 1.5 mL microtubes. Centrifuge for 1 min at $1000 \times g$ and discard flow-through.

16. Wash as in **step 5**, collecting the flow-through.

17. Cap the columns, and add 500 µL PBS. Optional pause point: Add sodium azide to a final concentration of 0.01% for long-term storage at 4 °C.

18. Immediately prior to serum purification, elute any nonspecifically bound proteins from the MBP-*Eh*MIF columns with glycine. Transfer the MBP-*Eh*MIF bound to NHS resin to a 10 mL column, rinsing with PBS to ensure all resin is transferred. Wash with 10 mL PBS to remove sodium azide if necessary.

19. Add 5 mL 0.1 M glycine pH 2.8 to the column and remove via gravity flow.

20. Wash column with 10 mL 1× PBS and collect via gravity flow. Repeat once to ensure all glycine is removed.

21. Add serum to column containing MBP bound to amylose resin. Rotate end-over-end for 1 h at room temperature or overnight at 4 °C.

22. Collect serum from MBP column via gravity flow, adding PBS until resin returns to its original color.

23. Add MBP-cleared serum to MBP-EhMIF column.

24. Rotate column end-over-end or mix for 2–3 h at room temperature or overnight at 4 °C. Collect serum via gravity flow, adding PBS until resin returns to its original color.

25. Elute antibodies with 1 mL 0.1 M glycine (pH 2.8) for every 2 mL serum originally added (*see* **Note 4**). Collect in 0.5–1 mL aliquots, and promptly neutralize aliquots with 1:10 1 M Tris–HCl, pH 9 (100 µL in each 1 mL aliquot immediately after collection).

26. Repeat glycine elution once.

27. Wash column with 10 mL PBS to ensure all glycine is removed and collect wash. Repeat as necessary. Column may be stored at 4 °C in PBS.

28. Anti-EhMIF dialysis may be completed as in Subheading 3.2, **step 27**.

3.4 ELISA to Detect EhMIF in Stool (Fig. 2a)

1. Coat 96-well high-protein-binding polystyrene plates with 5 µg/mL anti-*Eh*MIF in PBS overnight.

2. Block plate for 1 h with 1% (w/v) bovine serum albumin (BSA) in PBS.

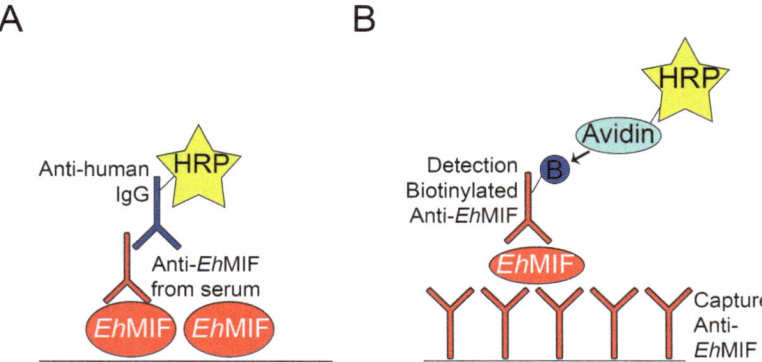

Fig. 2 Schematic representation of detection of anti-EhMIF antibody (**a**) and EhMIF protein using ELISA (**b**)

3. Using recombinant *Eh*MIF as a standard, incubate stool samples overnight at 4 °C.

4. Wash plate with PBS, pH 7.4 + 0.05% Tween-20, incubating plate for 1 min during wash.

5. Repeat wash for a total of 3–6 times until as much stool as possible is removed.

6. Add biotinylated anti-*Eh*MIF (*see* **Note** 7) at 0.25 μg/mL, diluted in 1% BSA in PBS, and incubate for 2 h.

7. Wash plate as in **step 4**, repeating for a total of three times.

8. Add avidin-conjugated horseradish peroxidase and incubate for 30 min.

9. Wash plate 5–7 times as in **step 4**.

10. Detect with 200 μL TMB ELISA detection reagent for 15 min, and stop reaction with 50 μL 1 M phosphoric acid.

11. Measure plate absorbance at 450 nm [2].

3.5 Serum Anti-EhMIF Detection-(Fig. 2b)

1. Coat a 96-well ELISA plate with recombinant *Eh*MIF at 0.5 μg per well in carbonate-bicarbonate buffer, pH 9.5. Incubate plate overnight at 4 °C.

2. Wash plate with PBS with 0.05% Tween-20 three times.

3. Block with 0.25% BSA in PBS-0.05% Tween-20 for 2 h at room temperature before washing again as in **step 2**.

4. Optionally, air-dry plate and store at 4 °C with desiccant for later use as a pre-coated plate.

5. Add 100 μL of 1:100 to 1:250 serum samples diluted in PBS, each sample in triplicate. Incubate samples overnight at 4 °C.

6. Wash plate three times as in **step 2**.

7. Add HRP-conjugated anti-human total IgG diluted 1:10,000 in 0.25% BSA in PBS-0.05% Tween-20 and incubate for 2 h.

8. Wash plate 5–7 times as in **step 2**, soaking plate for 1 min during washes.

9. Add 200 μL TMB ELISA detection reagent for 15 min and stop reaction with 50 μL 1 M phosphoric acid.

10. Measure plate absorbance at 450 nm [2].

11. Apply the following formula to determine sera positivity: $2 \times SD$ + average of negative (healthy, non-reactive) controls. Subtract the result from the average value of each sample; a positive number indicates positive sera while a negative or zero value indicates negative sera [10].

3.6 Immunohisto-chemistry (See Note 8)

1. Cut 4 μm sections from paraffin-embedded tissue and place on charged glass slides.

2. Deparaffinize slides and retrieve antigens in Target Retrieval Solution, high pH at 97 °C for 20 min, using a PT Link instrument.

3. The following steps are performed with a Dako Autostainer: Block endogenous peroxidases with Peroxidase and alkaline phosphatase blocking reagent.

4. Dilute anti-*Eh*MIF 1:600 in Dako Antibody Diluent, ready-to-use diluent, and apply to slides at ambient temperature for 30 min.

5. To visualize antibody binding, incubate slides with Envision Rabbit Link followed by 3, 3′-diaminobenzidine tetrachloride (DAB+).

6. Counterstain slides with hematoxylin.

7. Dehydrate, clear, and mount slides for assessment [4].

4 Notes

1. This protocol was used to purify rabbit anti-*Eh*MIF generated from recombinant His-tagged *Eh*MIF. If generating an antibody from a GST-tagged fusion protein, other high-affinity, high solubility fusion proteins can be used for purification, such as MBP [5]. Untagged proteins may also be used. For instructions regarding purification with MBP proteins using covalent coupling, *see* Subheadings 2.3 and 3.3. For untagged proteins, skip to **step 8** of Subheading 3.2 for Buffer Exchange into PBS or **step 10** if already in a non-amine-containing buffer.

2. You may also use any other non-amine-containing buffer compatible with glutathione sepharose, per the manufacturer's guidelines. Use a volume of 1 mL buffer to 100 mL original culture volume. Use the same non-amine-containing buffer for the rest of Subheading 3.2.

3. Optional pause point: serum may be 0.45 μm filtered and stored at 4 °C until GST-Fusion protein has been cross-linked to glutathione sepharose, washed, and cleaned of nonspecifically bound proteins. If proceeding directly from GST clearance to affinity purification, serum may also be transferred from the GST column to the GST-fusion protein column via gravity flow through use of ring stand and clamps.

4. Due to possible denaturation effects, do not incubate glycine elution; add glycine and collect immediately via gravity flow.

5. The post-elution PBS wash may also be included in dialysis as it may contain trace amounts of antibodies.

6. Buffer exchange into PBS reduces the Tris in amylose column buffer to μM concentration. For optimal binding efficiency, proteins should be in a non-amine-containing buffer for covalent coupling with NHS-activated agarose.

7. When used as a detection antibody, anti-*Eh*MIF (and any other antibody) should be biotinylated for an efficacious ELISA development. The EZ-Link Sulfo-NHS-LC-Biotinylation Kit (Thermo Fisher) quickly and effectively biotinylates the antibody when following the manufacturer's directions. Biotinylation can be validated via western blot against recombinant protein (used for ELISA standard) with anti-biotin HRP (Sigma, 1:10,000 in casein buffer) as a secondary.

8. The following procedure is performed on paraffin-embedded tissue for HRP development, but may be adapted to frozen sections and immunofluorescence.

References

1. Ghosh S et al (2018) Interaction between parasite-encoded JAB1/CSN5 and macrophage migration inhibitory factor proteins attenuates its proinflammatory function. Sci Rep 8(1):10241

2. Ngobeni R et al (2017) Entamoeba histolytica-encoded homolog of macrophage migration inhibitory factor contributes to mucosal inflammation during amebic colitis. J Infect Dis 215(8):1294–1302

3. Moonah SN et al (2014) The macrophage migration inhibitory factor homolog of Entamoeba histolytica binds to and immunomodulates host macrophages. Infect Immun 82(9):3523–3530

4. Shirley D-AT et al (2018) A review of the global burden, new diagnostics, and current therapeutics for amebiasis. Open Forum Infect Dis 5(7):ofy161

5. Moonah S et al (2014) Erythrocyte lysis and Xenopus laevis oocyte rupture by recombinant Plasmodium falciparum hemolysin III. Eukaryot Cell 13(10):1337–1345

6. Mori K et al (2013) The C9orf72 GGGGCC repeat is translated into aggregating dipeptide-repeat proteins in FTLD/ALS. Science 339(6125):1335–1338

7. Annis MG et al (2001) Endoplasmic reticulum localized Bcl-2 prevents apoptosis when redistribution of cytochrome c is a late event. Oncogene 20:1939

8. Watanabe K et al (2017) Microbiome-mediated neutrophil recruitment via CXCR2 and protection from amebic colitis. PLoS Pathog 13(8):e1006513

9. Schlaepfer DD et al (1994) Integrin-mediated signal transduction linked to Ras pathway by GRB2 binding to focal adhesion kinase. Nature 372(22):786–791

10. González-Canto A, et al (2017) Detection of specific antibodies for amoebapore in serum of patients with amoebic liver abscess. Ann Clin Pathol 5(1):1103–1110

Chapter 22

Studying Plant MIF/D-DT-Like Genes and Proteins (MDLs)

Dzmitry Sinitski, Katrin Gruner, Jürgen Bernhagen, and Ralph Panstruga

Abstract

Human macrophage migration inhibitory factor (MIF) is an inflammatory cytokine with chemokine-like characteristics and an upstream regulator of host innate immunity. It is a critical mediator of a variety of human diseases, such as acute and chronic inflammatory diseases, autoimmunity, atherosclerosis, and cancer. MIF is an atypical chemokine that not only signals through its cognate receptor CD74, but also interacts with the classical chemokine receptors CXCR2 and CXCR4. MIF and its homolog D-dopachrome tautomerase (D-DT)/MIF-2 are structurally unique proteins that are conserved across kingdoms and that share a remarkable homology with bacterial tautomerases/isomerases, albeit the relevance of the tautomerase activity in mammalian systems has remained unclear. Intriguingly, in silico analysis also predicts MIF orthologs in plants such as in the model plant *Arabidopsis thaliana*. There are three predicted MIF orthologs in *A. thaliana*, which have been termed *A. thaliana* MIF/D-DT-like proteins (*At*MDLs). Anticipating that there will be a future research interest in studying *At*MDLs or other plant MDLs, here we describe methods how to clone, recombinantly express and purify *At*MDL proteins, taking into account codon usage differences between plant and mammalian cell systems.

Key words *Arabidopsis thaliana*, *Arabidopsis thaliana* MIF/D-DT-like protein (*At*MDL), Chemotaxis, Cross-kingdom biology, Macrophage migration inhibitory factor (MIF), Innate immunity, Inflammation

1 Introduction

Human macrophage migration inhibitory factor (MIF) is an inflammatory cytokine with chemokine-like characteristics and is an upstream regulator of host innate immunity. Numerous studies in experimental disease models in rodents and a number of mostly observational clinical studies have identified MIF as a critical mediator of a variety of human diseases, such as acute and chronic inflammatory diseases, autoimmunity, atherosclerosis, and cancer [1–5]. MIF also is a prototypical member of the emerging class of atypical chemokines (ACKs). Thus, it not only binds to its cognate receptor CD74 [6], but also engages in high-affinity binding with the chemokine receptors CXCR2 and CXCR4 [7]. Human MIF is a

James Harris and Eric F. Morand (eds.), *Macrophage Migration Inhibitory Factor: Methods and Protocols*,
Methods in Molecular Biology, vol. 2080, https://doi.org/10.1007/978-1-4939-9936-1_22,
© Springer Science+Business Media, LLC, part of Springer Nature 2020

structurally unique 12.5 kDa protein and the founding member of the MIF protein family that also comprises D-dopachrome tautomerase (D-DT)/MIF-2 and MIF orthologs in various species [1, 7–11]. The sequence and three-dimensional (3D) structure of MIF proteins differs from that of other cytokines and classical chemokines [12], although a remote architectural similarity has been noted between MIF and CXCL8 [7, 12]. Interestingly, despite an only approximately 30% identity in amino acid sequence, mammalian MIF proteins share a remarkably high 3D-structural homology with bacterial tautomerases/isomerases [1, 13–16]. The acronym D-DT stems from the ability of this MIF homolog to convert the nonnatural D-stereoisomer of dopachrome into its L-tautomer. MIF shares this catalytic property with D-DT and can also tautomerize hydroxy-phenylpyruvate (HPP) [1, 14, 17]. Accordingly, MIF proteins can act as both cytokines or chemokines and enzymes, although the functional significance of the tautomerase activity in mammalians has remained unknown and physiological substrates have yet to be identified.

MIF proteins are highly conserved across kingdoms, ranging from mammalians to unicellular parasites [1, 11, 18]. As they also are expressed as cytosolic proteins that are released by a non-endoplasmic reticulum (ER)/Golgi-mediated unconventional secretion mechanism, MIF proteins might have been evolutionary old cytosolic enzymes that have adopted a secondary role as extracellular cytokines during evolution.

To the very best of our knowledge, the existence of *MIF*-like genes in plants was first reported in 1998 when a single MIF protein sequence of the model plant *Arabidopsis thaliana* was included in a multiple sequence alignment [19]. Later, the existence of three *MIF*-like genes in the genome of *A. thaliana* was realized, and their incorporation into a multi-species MIF phylogeny revealed that they cluster as a sister group of apicomplexan MIF sequences in a clade that is separate from vertebrate and non-vertebrate MIF, animal D-DT/MIF-2 and bacterial sequences [20]. Our own studies revealed the first comprehensive in silico analysis of the phylogeny and evolution of plant macrophage migration inhibitory factor/ D-dopachrome tautomerase-like proteins (MDLs). We found that MIF-like proteins are found in all species across the plant kingdom, with an increasing family complexity toward evolutionarily advanced plant *taxa* [21]. Conservation of MDLs throughout the plant kingdom suggests important function(s) of these proteins in plant life. However, to date no functional data exist for any plant MDL gene or protein.

As the majority of MIF researchers have so far studied mammalian MIF genes and proteins, available methods all relate to the specific needs and specialities of mammalian systems. These investigators are typically not familiar with the specific necessities of plant systems, ranging from cloning procedures to methods related to the expression, purification, and examination of plant MDLs. In an attempt to begin to experimentally study plant *MIF*-like genes and proteins, we here describe cloning, expression, and purification protocols for plant MDLs.

Cloning starts from the cDNA of *A. thaliana* plant tissue or from that of another plant of interest. If available, purified plasmid DNA containing the plant *MDL* gene of interest may be used as starting material. If plant *MDLs* are to be studied in mammalian cell systems for comparative reasons, expression plasmids with dual utility should be used (see below). In general, considerations to clone and study plant MDLs also should account for codon usage differences between mammalian, bacterial, and plant cell systems, as codon usage is a critical factor for heterologous gene expression, e.g., in microbial systems [22, 23]. Codon usage varies widely within genomic DNA sequences of different organisms, including plants. Early studies revealed that codon usage patterns in plants are more similar to man and other higher eukaryotic organisms than to unicellular organisms. This is primarily due to the preference for guanine and cytosine in the third codon position [24]. The peculiarities of plant codon usage may restrict heterologous gene expression (including expression of *MDL* genes) in bacterial systems such as *Escherichia coli*. Initially, it was thought that primarily the existence of rare codons may limit the translation of transgenes in heterologous organisms. There are two principal approaches to address this challenge. One possibility is to generate a codon-optimized version of the transgene [25], and the other option is to deploy bacterial strains that express rare t-RNAs to compensate for the codon bias [26]. With the advent of gene synthesis, codon optimization has widely become the prime method of choice. Several online tools are now available that assist researchers in gene design to improve codon usage for heterologous gene expression [27–29]. Indeed, similar to human genes [30], codon optimization of plant genes has been found to result in higher expression levels [31]. However, apart from affecting translation, codon usage has also been reported to limit transcription [32], mRNA levels [33] and secondary structure [34], initiation of translation [35], and co-translational protein folding [36]. Newly developed techniques aim to overcome also these shortcomings, e.g., by the prediction of context-specific codons [37] or usage of bicistronic design elements [38]. Depending on the particular plant *MDL* gene, one or several of the abovementioned methods might be deployed to improve heterologous expression and the production of sufficient levels of recombinant protein.

2   Materials

We recommend using freshly prepared or aliquoted and non-expired solutions only. Unless indicated otherwise, solutions and materials should be stored and handled as recommended by the manufacturer. In general, solutions should be prepared under sterile conditions using double-distilled water. Waste disposal rules as per laboratory and authority regulations must be followed for all used materials, when disposing of waste materials.

2.1   General

1. Biosafety level 1 laboratory for work with BSL 1 genetically modified organisms.
2. High- and medium-speed centrifuges.
3. Autoclave.
4. Spectrophotometer for UV-VIS measurements.
5. Water bath, 37 °C.
6. Standard molecular biology equipment.
7. Thermocycler for polymerase chain reactions (PCRs).
8. Reaction tubes: 0.5 mL, 1.5 mL, 15 mL, and 50 mL.
9. French Press.

2.2   Cloning of MDL Proteins from Arabidopsis thaliana

1. cDNA from plant tissue or purified plasmid DNA containing the plant *MDL* gene of interest.
2. Components of a standard PCR.
3. Oligonucleotide primers as required (see description below).
4. pET-21a(+) vector with an C-terminal hexahistidine-tag sequence.
5. Conventional DNA purification/concentration kits.
6. 1% Agarose gel supplemented with a DNA dye (e.g., ethidium bromide at a concentration of 500 ng/mL).
7. DNA loading dye.
8. DNA ladder with appropriate length spectrum.
9. *Nde*I and *Xho*I restriction enzymes with the appropriate buffer.
10. T4 DNA ligase with a 10× T4 DNA ligase buffer as provided together with the enzyme by the supplier.
11. Competent *E. coli* Top10 or DH5α strains for multiplication.
12. Competent *E. coli* Rosetta™ or BL21 strains for expression.
13. Restriction enzymes with the appropriate buffer as required (see description below).
14. Glycerol.

2.3 Expression of Recombinant A. thaliana MDL (AtMDL) Proteins

1. Luria Bertani (LB) growth medium: 25 g of LB broth in 1 L ddH$_2$O. Autoclave and store at 4 °C. If needed, add ampicillin/chloramphenicol at a final concentration of 100 μg/mL and 34 μg/mL, respectively.

2. LB agar plates: Prepare 1% LB/agar medium solution. Autoclave and pour the melted solution under sterile conditions into Petri dishes and wait until agar will solidify. If needed, add ampicillin/chloramphenicol at a final concentration of 100 μg/mL and 34 μg/mL, respectively, when the solution has not fully cooled down and is at around 50 °C.

3. A 500 mL baffled flask containing 250 mL of LB medium pre-warmed to 37 °C.

4. Isopropyl-β-D-1-thiogalactopyranoside (IPTG, 1 mM) for inducing the transgene expression.

2.4 Purification of Recombinant A. thaliana MDL (AtMDL) Proteins

1. Avestin Emulsiflex-C5 homogenizer, pre-washed with 20% ethanol/ddH$_2$O and equilibrated in immobilized metal ion affinity chromatography (IMAC) binding buffer (composition see below).

2. Fast protein liquid chromatography system (FPLC, ÄKTA Pure, GE Healthcare) washed and equilibrated with IMAC binding buffer.

3. High-speed centrifuge (e.g., Avanti JNX-26, Beckman Coulter).

4. 1 mL His-Trap columns (for IMAC), pre-equilibrated in binding buffer in accordance with the manufacturer's protocol (GE Healthcare).

5. Superdex™ 75 10/300 GL column for size exclusion chromatography (SEC), washed and equilibrated in 20 mM sodium phosphate buffer.

6. Buffers:

   (a) *IMAC binding buffer:* 20 mM sodium phosphate buffer, pH 7.2, 0.5 M NaCl, 20 mM imidazole.

   (b) *IMAC elution buffer:* 20 mM sodium phosphate buffer, pH 7.2, 0.5 M NaCl, 500 mM imidazole.

   (c) *Working and SEC elution buffer:* 20 mM sodium phosphate buffer, pH 7.2.

7. Syringes and sterile filtering device (pore size: 0.2 μm).

8. Sodium dodecyl-sulfate polyacrylamide gel electrophoresis (SDS-PAGE): 15% SDS polyacrylamide gels and PAGE running buffer (25 mM Tris, 0.2 M glycine, 0.5% SDS, pH 8.6).

3   Methods

3.1   Cloning of MDL Proteins from Arabidopsis Thaliana

3.1.1   Classical Cloning Strategy

1. Check the sequence of your desired plant *MDL* gene for internal *Nde*I and *Xho*I restriction sites. If the gene is clear of either restriction site, such as in case of *AtMDL2* (*see* Fig. 1), continue with **step 3**. If the gene contains one of the restriction sites, such as in cases of *AtMDL1* and *AtMDL3*, continue with **step 2**.

2. Apply splice overlap extension PCR (SOE-PCR; *see* **Note 1**) strategy to remove unwanted internal *Nde*I and *Xho*I restriction sites, such as the internal *Nde*I site in *AtMDL1* and *AtMDL3*, by introducing a silent mutation. Withal, use the mutagenesis-introducing internal primers "gene"-mut-Fwd and "gene"-mut-Rev in combination with the restriction site-adding primers "gene"-*Xho*I-Rev and *Nde*I-"gene"-Fwd (Table 1). The SOE-PCR modifies the cDNA sequences of *AtMDL1* and *AtMDL3* at position 177 (GCA to GCG) by introducing a silent mutation (A59A).

3. Carry out a standard/mutagenesis PCR on your desired plant *MDL* gene, e.g., *AtMDL2*, using plant tissue-derived cDNA or a plasmid containing the plant *MDL* gene of interest as template DNA and a primer pair that adds the restriction sites *Nde*I to the 5′ end and *Xho*I to the 3′ end of the gene (*see* **Notes**

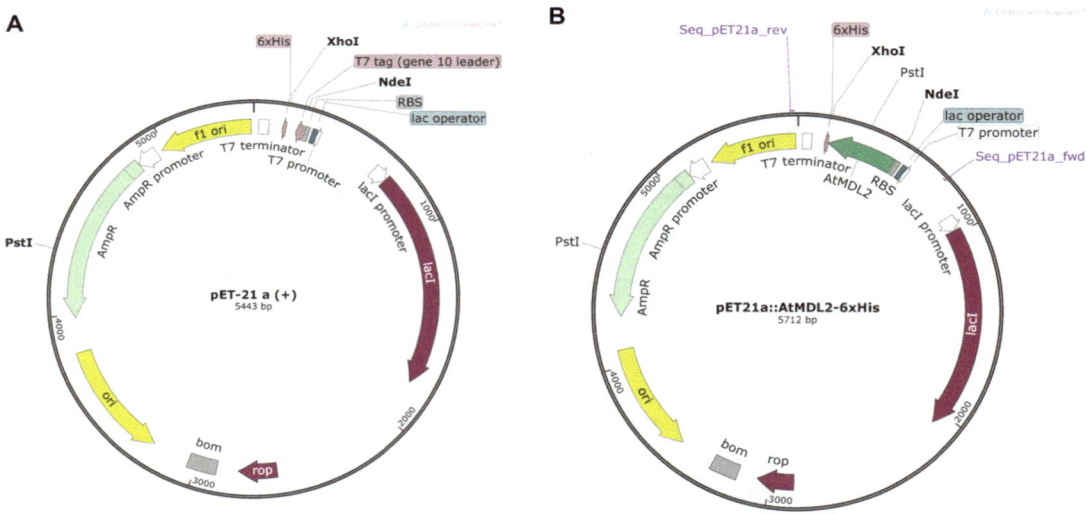

Fig. 1 pET-21a(+) and pET-21a::*At*MDL2-6xHis vectors. (**a**) The original pET-21a(+) vector. Indicated are the *Nde*I and *Xho*I restriction sites for opening of the vector upstream of the 6xHis-tag, and the unique *Pst*I restriction site in the backbone. (**b**) pET-21a::*At*MDL2-6xHis is displayed as an example. *At*MDL2 was inserted via *Nde*I and *Xho*I restriction sites, and the gene was C-terminally fused with the 6xHis-tag. Further indicated are two *Pst*I restriction sites, one in the backbone, one in *At*MDL2, and the binding sites of the sequencing primers. Plasmid maps were created with SnapGene

Table 1
Primers used in (SOE-)PCR adding *Nde*I and *Xho*I restriction sites to *AtMDL1*, *AtMDL2*, and *AtMDL3*

Target gene	NCBI Ref. Seq.	Primer name	Primer sequence (5′ to 3′)
AtMDL1 (At5g57170.1)	NM_125099.4	*Nde*I-*AtMDL1*_Fwd	CGG**CATATG**CCCACTTTGAATCTC
		AtMDL1-*Xho*I_Rev	CGG**CTCGAG**GAAAGTTGATCCATTG TAACC
		*AtMDL1*_mut_Fwd	GAGGAACCTGCTGCGTATGGAGAA TTGATATC
		*AtMDL1*_mut_Rev	GATATCAATTCTCCATACGCAGCAGG TTCCTC
AtMDL2 (At5g01650.1)	NM_120243.4	*Nde*I-*AtMDL2*_Fwd	GAA**CATATG**CCGTGCCTCAACCTC TCC
		AtMDL2-*Xho*I_Rev	CAA**CTCGAG**AAGAGTCGCCCCG TTCCAACC
AtMDL3 (At3g51660.1)	NM_115025.4	*Nde*I-*AtMDL3*_Fwd	GCC**CATATG**CCTTGTCTTTACATTAC
		AtMDL3-*Xho*I_Rev	CGG**CTCGAG**AAG TTTAGAAGGAAGAGGCAAAG
		*AtMDL3*_mut_Fwd	CAAAGAAGCAGCTGCGTA TGCAGAGATTGTGTC
		*AtMDL3*_mut_Rev	GACACAATCTCTGCATACGCAGC TGCTTCTTTG

Indicated in bold are the restriction sites

2 and **3**). The reverse primer at the same time serves to remove the endogenous stop codon. Table 1 exemplifies the restriction site-adding primer pair *Nde*I-"gene"-Fwd/"gene"-*Xho*I-Rev for *AtMDL1*, *AtMDL2*, and *AtMDL3*.

4. Purify the (SOE-)PCR product (*see* **Note 4**).

5. Verify successful amplification *via* the expected size of the "*Nde*I-*plantMDL*-*Xho*I" DNA fragment by separating a small amount of the purified PCR product (e.g., 5 μL from a total of 25 μL) on an agarose gel supplemented with a DNA dye.

6. Digest the purified PCR product with *Nde*I and *Xho*I.

7. Purify the digested fragment (*see* **Note 4**).

8. Open the hexahistidine-containing (6xHis) pET-21a vector by digesting 1 μg plasmid DNA with *Nde*I and *Xho*I (Fig. 1a). *Xho*I opens the vector right upstream of the 6xHis.

9. Separate the digested pET-21a vector on an agarose gel supplemented with a DNA dye and purify the expected band (digested **5363** bp *versus* **5443** bp not digested) (*see* **Note 4**).

10. Estimate the amount of the resulting digested fragment and vector by running a small amount of each purified product

(e.g., 5 μL from a total of 25 μL) on an agarose gel supplemented with a DNA dye.

11. Ligate the opened vector with the digested insert fragment in an approximately 1:3 ratio, respectively, to the desired pET-21a::*plantMDL*-6xHis product (Fig. 1b).

12. Transform the resulting pET-21a::*plantMDL*-6xHis ligation product for propagation into competent *E. coli* Top10 or DH5a cells and plate them on selective ampicillin-containing LB agar. Incubate overnight at 37 °C.

13. The next day, pick grown colonies to inoculate up to 6 mL liquid LB cultures containing ampicillin. Incubate the cultures overnight at 37 °C shaking (ca. 220 rpm).

14. The next day, make minipreps from 4 mL of the cultures to isolate the plasmid DNA (*see* **Note 4**). Keep 2 mL of the cultures for possible glycerol stock preparation.

15. Digest minipreps using appropriate restriction enzymes that allow to discriminate positive clones exhibiting the desired ligation constructs from colonies lacking the desired ligation constructs (*see* **Note 5**). As a negative control, also digest the original pET-21a vector. For example: *Pst*I cuts once in the backbone of pET-21a, resulting in a 5443 bp fragment in case of the original plasmid without an insert. A second *Pst*I restriction site is located in the *AtMDL2* gene, resulting in two fragments of 1478 bp and 4234 bp in case of a successful pET-21a::*AtMDL2*-6xHis ligation product.

16. Separate the digested plasmid DNA on an agarose gel supplemented with a DNA dye and check for expected band sizes.

17. Sequence the clones with expected band pattern to verify the correct insertion of the plant *MDL* gene in the pET-21a vector in frame with the 6xHis-tag at the 3' end. Sequencing primer forward: 5'-TACCCACGCCGAAACAAG-3', and reverse: 5'-TTAATGCGCCGCTACAGG-3'.

3.1.2 Competent Bacteria and Storage

1. Retransform verified pET-21a::*plantMDL*-6xHis constructs into competent *E. coli* Rosetta™ or BL21 cells for protein expression purposes.

2. For long-term storage at −80 °C, prepare glycerol stocks of verified *E. coli* carrying the desired pET-21a::*plantMDL*-6xHis plasmid.

3.2 Bacterial Expression of MDL Proteins from Arabidopsis Thaliana

1. Prepare LB medium and LB agar Petri dishes with ampicillin/chloramphenicol as described in section 2.3 in advance.

2. Plate the transformed bacteria with one of the recombinant *MDL* genes on LB agar plates with ampicillin/chloramphenicol 2 days before starting the experiment. Incubate at 37 °C

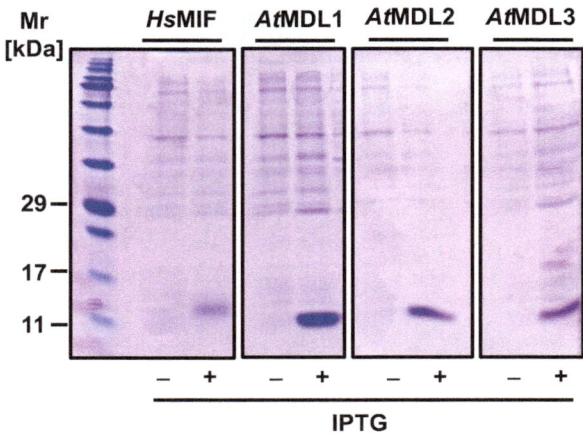

Fig. 2 Coomassie brilliant blue staining of a polyacrylamide gel after SDS-PAGE analysis of lysed bacterial pellets from bacterial cultures expressing hexahistidine-tagged variants of the three *At*MDLs *At*MDL1, *At*MDL2, and *At*MDL3 before and after induction of protein expression by IPTG. The expression of human MIF (*Hs*MIF) was electrophoresed for comparison

overnight. In the second half of the next day, inoculate 5 mL of LB medium with a single colony from the plate and incubate on the shaker (220 rpm) at 37 °C overnight.

3. On the next day, take 1 mL of bacterial culture to inoculate 250 mL of fresh pre-warmed LB medium in a 500 mL Erlenmeyer flask and continue the incubation on the shaker at 37 °C with the same parameters.

4. After 3 h of incubation, check the optical density of the bacterial cultures using spectrophotometry at a wavelength of 600 nm (OD_{600}). When the OD_{600} reaches 0.5–0.8, add the freshly aliquoted IPTG solution to the culture to a final concentration of 1 mM and continue the incubation for another 3 h.

5. After the incubation, collect the bacterial culture and centrifuge it at $5000 \times g$ in a cooled centrifuge (at 4 °C) for 30 min.

6. Discard the supernatant and proceed to bacterial lysis or storage by freezing the pellets at −20 °C for later processing (stop **step 1**) (*see* **Note 6**).

7. Verify expression of the target protein by SDS-PAGE / Coomassie staining (Fig. 2).

3.3 Purification of AtMDL Proteins

1. Figure 3 illustrates the purification workflow.

2. Prepare the FPLC system and the His-Trap™ and Superdex™ 75 10/300 GL columns in accordance with the manufacturer's instructions.

Fig. 3 Workflow of *Arabidopsis thaliana* MDL (*At*MDL) and human MIF expression and purification procedures

3. Thaw the bacteria pellets on ice and resuspend them in binding buffer (approx. 2 mL of buffer per pellet from a 50 mL bacterial culture). Using a syringe, pass the suspension 5–10 times through a 0.7 mm needle. Lyse the bacteria using a French press.

4. Centrifuge the collected bacterial lysate at 39,000 × *g* and 4 °C for 30 min. Save the supernatant for further purification, but do not throw away the pellet, in case your target protein partitions into inclusion bodies.

5. Check the localization of the recombinant MDL protein by running an SDS-PAGE analysis. To do so, add the supernatant and/or pellet to lithium dodecyl-sulfate/dithiothreitol (LDS/DTT) lysis buffer, boil the mixtures, and perform the SDS-PAGE according to routine Laemmli procedures (*see* **Notes 7** and **8**).

6. After confirmation of the solubility of the MDL protein and its partitioning into the supernatant, proceed with IMAC purification (*see* **Note 9**). Keep the protein supernatants on ice until further processing.

7. Load the total supernatant fraction onto the IMAC column applying constant flow of binding buffer (i.e., 1 mL/min). After elution of those proteins that do not bind to the column, run a gradient of elution buffer up to 100% over 20 min. Collect the eluted fractions; an increase in UV absorption at 280 nm indicates the elution of hexahistidine-tagged MDL protein. Verify successful elution of the target protein by SDS-PAGE in combination with Coomassie brilliant blue and/or silver staining (*see* **Note 10**). IMAC typically results in 80–90% pure target protein.

8. After the IMAC purification step, proceed with size exclusion chromatography (SEC) to purify the protein further. SEC also serves to exchange the working buffer (i.e., to 20 mM sodium phosphate, pH 7.2, needed for functional assays). Equilibrate the Superdex™ 75 10/300 GL column with 20 mM sodium phosphate, pH 7.2, and afterward load the column with the IMAC-purified MDL fractions. Elute in 20 mM sodium phosphate, pH 7.2. Collect OD_{280}-positive fractions as described above and verify protein identity and purity by SDS-PAGE in combination with Coomassie brilliant blue and/or silver staining (*see* **Note 11**).

9. After final confirmation of target protein purity by SDS-PAGE, measure the protein concentration using Bradford assay or bicinchoninic assay (BCA) (*see* **Note 12**).

10. Store the protein in 20 mM sodium phosphate buffer, pH 7.2, at 4 °C until further use; it is recommended to perform functional experiments within 4 weeks (*see* **Note 13**).

4   Notes

1. Gene splicing by overlap extension is described in detail by Horton et al., 1993 [39].

2. The *Nde*I restriction site already includes the ATG-start codon.

3. For cleavage close to the end of a DNA fragment, to add a sufficient number of bases to the ends of the PCR primers to allow for efficient restriction enzyme activity.

4. For DNA purification and concentration after PCR, restriction digest, or from agarose gels supplemented with a DNA dye, a variety of kits as well as standard laboratory methods are available.

5. Alternatively, colony PCR can be applied to identify clones exhibiting the desired ligation product.

6. During incubation of the bacteria and before the purification procedure, take an aliquot (i.e., 100 μL) of culture liquid just before the induction, as well as one after IPTG induction. Centrifuge these culture aliquots at $5000 \times g$ for 5 min, and lyse the cell pellet in LDS/DTT ("Laemmli") lysis buffer, and run an SDS-PAGE to confirm the successful induction of the recombinant target protein expression (for an example *see* Fig. 2).

7. SDS-PAGE is performed using a 15% acrylamide gel and an electrophoresis running program of 120 V for 1.5 h.

8. Upon lysis of the *E. coli Rosetta* strain expressing *Arabidopsis thaliana*, hexahistidine-tagged MDL proteins and *At*MDL2 are fully soluble, but *At*MDL3 has a tendency to aggregate in inclusion bodies. Therefore, *At*MDL3 yields are lower. To obtain higher yields for *At*MDL3, inclusion bodies need to solubilized by chaotropic reagents. This requires subsequent refolding of the target protein.

9. All buffers used for chromatography should be degased and sterile-filtered before usage. Total protein after pellet lysis should be filtered using a 0.2 μm filter device.

10. After SDS-PAGE, if necessary, perform Western blot analysis to verify the enrichment of the hexahistidine-tagged MDL proteins using an anti-His antibody. For membrane transfer, use a

standard Western blot protocol for nitrocellulose membranes and a routine Western blot program (30 V power, 1.5 h).

11. After the IMAC and SEC purification steps, MDL proteins do not require to undergo a refolding procedure, as strong denaturants were not used.

12. If the protein concentration is unfavorable (i.e., too dilute or too concentrated) for the desired subsequent assay, dilute or concentrate the protein as appropriate.

13. Do not store *At*MDL proteins at concentrations exceeding 4–5 mg/mL, as they may precipitate.

Acknowledgments

This work was supported by the Deutsche Forschungsgemeinschaft (DFG)-Agence Nationale Recherche (ANR) co-funded project "X-KINGDOM-MIF - Cross-kingdom analysis of macrophage migration inhibitory factor (MIF) functions." Respective DFG grants are BE 1977/10-1 to J.B. and PA 861/15-1 to R.P. Additional funding was provided by the DFG under Germany's Excellence Strategy within the framework of the Munich Cluster for Systems Neurology EXC 2145 SyNergy [grant number 390857198].

References

1. Calandra T, Roger T (2003) Macrophage migration inhibitory factor: a regulator of innate immunity. Nat Rev Immunol 3 (10):791–800

2. Tillmann S, Bernhagen J, Noels H (2013) Arrest functions of the mif ligand/receptor axes in atherogenesis. Front Immunol 4:115

3. Morand EF, Leech M, Bernhagen J (2006) MIF: a new cytokine link between rheumatoid arthritis and atherosclerosis. Nat Rev Drug Discov 5(5):399–410

4. Zernecke A, Bernhagen J, Weber C (2008) Macrophage migration inhibitory factor in cardiovascular disease. Circulation 117 (12):1594–1602

5. Sinitski D et al (2019) Macrophage Migration inhibitory factor (MIF)-based therapeutic concepts in atherosclerosis and inflammation. Thromb Haemost. https://doi.org/10.1055/s-0039-1677803

6. Leng L et al (2003) MIF signal transduction initiated by binding to CD74. J Exp Med 197 (11):1467–1476

7. Bernhagen J et al (2007) MIF is a noncognate ligand of CXC chemokine receptors in inflammatory and atherogenic cell recruitment. Nat Med 13(5):587–596

8. David JR (1966) Delayed hypersensitivity in vitro: its mediation by cell-free substances formed by lymphoid cell-antigen interaction. Proc Natl Acad Sci U S A 56(1):72–77

9. Bernhagen J et al (1993) MIF is a pituitary-derived cytokine that potentiates lethal endotoxaemia. Nature 365(6448):756–759

10. Merk M et al (2012) D-dopachrome tautomerase (D-DT or MIF-2): doubling the MIF cytokine family. Cytokine. https://doi.org/10.1016/j.cyto.2012.03.014

11. Bloom J, Sun S, Al-Abed Y (2016) MIF, a controversial cytokine: a review of structural features, challenges, and opportunities for drug development. Expert Opin Ther Targets 20(12):1463–1475

12. Sun HW et al (1996) Crystal structure at 2.6-A resolution of human macrophage migration inhibitory factor. Proc Natl Acad Sci U S A 93 (11):5191–5196

13. Kapurniotu A, Gokce O, Bernhagen J (2019) The multitasking potential of alarmins and atypical chemokines. Front Med (Lausanne) 6:3

14. Lolis E, Bucala R (2003) Macrophage migration inhibitory factor. Expert Opin Ther Targets 7(2):153–164

15. Stamps SL, Fitzgerald MC, Whitman CP (1998) Characterization of the role of the amino-terminal proline in the enzymatic activity catalyzed by macrophage migration inhibitory factor. Biochemistry 37 (28):10195–10202

16. Taylor AB et al (1999) Crystal structure of macrophage migration inhibitory factor complexed with (E)-2-fluoro-p-hydroxycinnamate at 1.8 A resolution: implications for enzymatic catalysis and inhibition. Biochemistry 38 (23):7444–7452

17. Merk M et al (2011) The D-dopachrome tautomerase (DDT) gene product is a cytokine and functional homolog of macrophage migration inhibitory factor (MIF). Proc Natl Acad Sci U S A 108(34):E577–E585

18. Sparkes A et al (2017) The non-mammalian MIF superfamily. Immunobiology 222 (3):473–482

19. Esumi N et al (1998) Conserved gene structure and genomic linkage for D-dopachrome tautomerase (DDT) and MIF. Mamm Genome 9(9):753–757

20. Miska KB et al (2007) Characterisation of macrophage migration inhibitory factor from Eimeria species infectious to chickens. Mol Biochem Parasitol 151(2):173–183

21. Panstruga R, Baumgarten K, Bernhagen J (2015) Phylogeny and evolution of plant macrophage migration inhibitory factor/D-dopachrome tautomerase-like proteins. BMC Evol Biol 15:64

22. Angov E (2011) Codon usage: nature's roadmap to expression and folding of proteins. Biotechnol J 6(6):650–659

23. Gustafsson C, Govindarajan S, Minshull J (2004) Codon bias and heterologous protein expression. Trends Biotechnol 22(7):346–353

24. Murray EE, Lotzer J, Eberle M (1989) Codon usage in plant genes. Nucleic Acids Res 17 (2):477–498

25. Elena C et al (2014) Expression of codon optimized genes in microbial systems: current industrial applications and perspectives. Front Microbiol 5:21

26. Lee SF, Li YJ, Halperin SA (2009) Overcoming codon-usage bias in heterologous protein expression in Streptococcus gordonii. Microbiology 155(Pt 11):3581–3588

27. Chin JX, Chung BK, Lee DY (2014) Codon Optimization OnLine (COOL): a web-based multi-objective optimization platform for synthetic gene design. Bioinformatics 30 (15):2210–2212

28. Puigbo P et al (2007) OPTIMIZER: a web server for optimizing the codon usage of DNA sequences. Nucleic Acids Res 35(Web Server issue):W126–W131

29. Grote A et al (2005) JCat: a novel tool to adapt codon usage of a target gene to its potential expression host. Nucleic Acids Res 33(Web Server issue):W526–W531

30. Burgess-Brown NA et al (2008) Codon optimization can improve expression of human genes in Escherichia coli: A multi-gene study. Protein Expr Purif 59(1):94–102

31. Xue F et al (2016) Expression of codon-optimized plant glycosyltransferase UGT72B14 in Escherichia coli enhances salidroside production. Biomed Res Int 2016:9845927

32. Zhou Z et al (2016) Codon usage is an important determinant of gene expression levels largely through its effects on transcription. Proc Natl Acad Sci U S A 113(41): E6117–E6125

33. Boel G et al (2016) Codon influence on protein expression in E. coli correlates with mRNA levels. Nature 529(7586):358–363

34. Zama M (1990) Codon usage and secondary structure of mRNA. Nucleic Acids Symp Ser 22:93–94

35. Bentele K et al (2013) Efficient translation initiation dictates codon usage at gene start. Mol Syst Biol 9:675

36. Buhr F et al (2016) Synonymous codons direct cotranslational folding toward different protein conformations. Mol Cell 61(3):341–351

37. Tian J et al (2017) Predicting synonymous codon usage and optimizing the heterologous gene for expression in E. coli. Sci Rep 7 (1):9926

38. Nieuwkoop T, Claassens NJ, van der Oost J (2019) Improved protein production and codon optimization analyses in Escherichia coli by bicistronic design. Microb Biotechnol 12(1):173–179

39. Horton RM et al (1993) Gene splicing by overlap extension. Methods Enzymol 217:270–279

INDEX

James Harris and Eric F. Morand (eds.), *Macrophage Migration Inhibitory Factor: Methods and Protocols*,
Methods in Molecular Biology, vol. 2080, https://doi.org/10.1007/978-1-4939-9936-1,
© Springer Science+Business Media, LLC, part of Springer Nature 2020